Geotechnical Problems and Solutions

Geotechnical Problems and Solutions

A Practical Perspective

Buddhima Indraratna, Ana Heitor
and Jayan S. Vinod

CRC Press
Taylor & Francis Group
Boca Raton London New York Leiden

CRC Press is an imprint of the
Taylor & Francis Group, an **informa** business

A BALKEMA BOOK

CRC Press/Balkema is an imprint of the Taylor & Francis Group, an informa business

© 2021 Taylor & Francis Group, London, UK Typeset by Apex CoVantage, LLC

Library of Congress Cataloging-in-Publication Data
Applied for

Published by: CRC Press/Balkema
                Schipholweg 107C, 2316 XC Leiden, The Netherlands
                e-mail: Pub.NL@taylorandfrancis.com
                www.crcpress.com – www.taylorandfrancis.com

ISBN: 978-1-138-48944-8 (hbk)
ISBN: 978-1-351-03734-1 (eBook)
DOI: 10.1201/9781351037341
https://doi.org/10.1201/9781351037341

Contents

Foreword

As real-life applications in geotechnical engineering have evolved rapidly over the past few decades, the need for becoming familiar with the extended concepts and methods of soil mechanics, as well as developing modern problem- solving skills, has become imperative for students and practitioners alike. Most simple problems in soil mechanics can be solved using fundamental principles. While these solutions may not be all-encompassing when compared with sophisticated computer- aided numerical analyses, the analytical training offered via this book is not only an immense help but also provides, for students, a custom- made guide to essential learning requirements. For practitioners, the contents of the book provide a most useful and concise refresher of the theoretical concepts, plus a means of carrying out preliminary design tasks and interim checks on the calculations and results obtained from more advanced computational modeling.

The book, through its 15 chapters, covers the majority of pertinent aspects of soil mechanics and geotechnical applications in a succinct manner. They include fundamental concepts of weight–volume relationships, effective stress principles, permeability and seepage, elastic stress analysis, foundations and retaining walls, slope stability analysis, critical state and yield criteria and unsaturated soil mechanics, plus an extended focus on applied topics including pavement engineering, excavation and tunneling, reclamation works and the use of alternative materials including granular wastes. A summary of concepts and their theoretical background is logically followed by worked- out examples that are intended to help the reader in a "tutorialstyle" environment. The authors are to be commended for presenting often difficult technical material in a concise style for optimum learning, and in this regard, the book offers both students and practitioners a most useful quick reference and a guide to essential problem solving. While it is not intended to replace standard soil mechanics textbooks that are more descriptive with detailed background review, the array of worked-out examples and the step-by-step methods of solution to a range of geotechnical problems is an obvious highlight of this book.

There is no doubt in my mind that the manner in which the technical content is presented herein and the nature of practical problems described with appropriate solutions amply reflect the authors' research experience in a wide array of geotechnical disciplines. I congratulate Distinguished Professor Buddhima Indraratna, Dr. Ana Heitor and Associate Professor Jayan S. Vinod for this concise, elegant and timely contribution to geotechnical engineering education, and in particular for their efforts in training young professionals to adapt to the modern day requirements of industry.

Prof. Harry G. Poulos
Senior Consultant, Coffey Services Australia,
and Emeritus Professor, University of Sydney

Preface

Principles of soil mechanics based on mathematics and fundamental physics have changed little over the years, but the type and nature of applications and the corresponding technical standards to meet modern-day demands have transformed swiftly and considerably while incorporating enhanced skills in computer-aided analysis, teamwork-inspired design and improved communication. Driven by environmental and socio-economic awareness, inevitable reforms have been made to foster sustainable infrastructure and cost-effective construction practices along with a favorable carbon footprint in relation to materials use and reuse. These have made a significant impact on modern-day tertiary education. This book is an attempt to highlight the essential principles of geomechanics and elucidate the pathways from theory to practice through selected problems of field applications with practical solutions, which are expected to benefit civil engineering students and practitioners alike.

This book is not intended to be duplicative of numerous textbooks in soil mechanics and geotechnical engineering. In preparing this book, the authors have recognized the wide array of textbooks already in circulation, and therefore, the theoretical and conceptual background is provided only succinctly at the beginning of each chapter with its stand-alone technical content, while the scope is distinctly attuned to the engineering solutions to the problems presented. In this regard, each chapter contains several selected worked-out examples, often representing real-life situations and demonstrating the applications of geotechnical theory and concepts.

At the outset, Chapter 1 offers the basics of soil properties and classification methods, followed by fundamentals of weight–volume relationships in Chapter 2. Then the most relevant subject matter pertaining to soil behavior including the effective stress principle and its applications, permeability and seepage, elastic stress analysis, shear strength criteria, foundations and retaining walls, stability of slopes and mass movement, excavations and tunneling, as well as key aspects of unsaturated soil mechanics are covered in the remaining chapters in a logical sequence. The book also contains a few specialized chapters on pavement engineering with applications to roads and rail corridors that offer the readers practical elements of transport infrastructure, which is still an evolving but a demanded field that now encompasses the utilization of waste materials. Critical state soil mechanics and corresponding stress paths with applications to practical situations are discussed where warranted.

Overall, the material provided in the standalone chapters has been developed over two decades of undergraduate and postgraduate teaching at the University of Wollongong, Australia. The contents have been inspired by numerous projects in which the authors have either been engaged as consultants or have undertaken contract research. In this regard, there are many colleagues and students both past and present to whom the authors are most grateful, as it was indeed their encouragement to produce a book such as this that had been the paramount impetus.

This Book, *Geotechnical Problems and Solutions-A Practical Perspective*, provides an ideal supplementary resource to be used in tandem with other comprehensive textbooks for an enhanced students' learning experience, hence a complementary and similar style of presentation has been maintained in most Chapters of the Book. In particular, numerous conceptual notes and illustrated examples contained in this Book have been truly inspired by authors of various other textbooks (multiple editions over the years) in the field of soil mechanics, foundations, and geotechnical engineering, now adopted worldwide in many tertiary education and research institutions. In this regard, during the preparation of this Book, we gratefully acknowledge the inspiration drawn and assistance through the past books of: Lamb, T. W. & Whitman, R.V., Terzaghi, K. & Peck, R.B., Scott, C. R., Das, B.M., Capper, P.L., Cassie, W.F. & Geddes, J.D., Craig, R.F., Smith, G.N., Sutton, B.H.C., Whitlow, R., Atkinson, J., Scott, R.F., Mitchell, J.K. & Soga, K., Holtz, R.D. & Kovacs, W.D., Terzaghi, K., Peck, R.B. & Mesri, G., Sivakugan, N. & Das, B.M., Briaud, J-L, Bowles, J.E., Smith, G.S. & Pole, E.L., Atkinson, J. & Bransby, P. L., Tomlinson, M J., Knappett, J. & Craig, R.F., Wood, D. M., Barnes, G.E. among others.

Special acknowledgments

The authors sincerely appreciate the constructive feedback of colleagues including Prof. Cholachat Rujikiatkamjorn, Dr. Trung Ngo, Dr. Yujie Qi, Dr. Chamindi Jayasuriya and Dr. Pankaj Baral. The immense help of a number of doctoral students who have been tutors in several geotechnical engineering subjects at the University of Wollongong is gratefully appreciated. In particular, Miriam Tawk, Mandeep Singh, Pubudu Jayathilaka, Aruni Abeygunewardena, Subhani Medewela, Soumyaranjan Mishra, Rakesh Mallisetty, Warranga Habaraduwa, Chathuri Arachchige, Senura Athuraliya, Fatima Mehmood, Marlisio Junior and Chuhao Liu, Ramesh Gadela, and Soumyaranjan Mishra, among others, have proofread the contents and assisted in the final artwork where necessary.

Buddhima Indraratna, Ana Heitor and Jayan S. Vinod
October 2019

About the authors

Distinguished Professor Buddhima Indraratna is a civil engineering graduate from Imperial College, London, and obtained his PhD from the University of Alberta in 1987. He has worked in industry in several countries before becoming an academic and has been a United Nations expert and a foreign advisor to numerous overseas projects. Prof. Indraratna's pioneering contributions to railway geotechnology and various aspects of geotechnical engineering have been acknowledged through many national and international awards, including the 1st Ralph Proctor Lecture and 4th Louis Menard Lecture of the International Society of Soil Mechanics and Geotechnical Engineering, ISSMGE, 2015 Thomas Telford Premium Award for Ground Improvement (ICE, UK), 2009 EH Davis Memorial Lecture of Australian Geomechanics Society, and 2017 Outstanding Contributions Medal of International Association for Computer Methods and Advances in Geotechnics, IACMAG, for his substantial and sustained contributions to Transport Geotechnics and Ground Improvement. Nationally, he was awarded the Engineers Australia Transport Medal in 2011, and more recently, the State Minister of Transport honored him with the 2015 Australasian Railway Technical Society's Outstanding Individual Award at the NSW Parliament. His pioneering contributions to railway engineering and ground improvement earned him the Fellowship of the Australian Academy of Technological Sciences and Engineering (FTSE) in 2011. Prof Indraratna is the Founding Director of Australian Research Council funded Industrial Transformation Training Centre, ITTC-Rail, and formerly Distinguished Professor of Civil Engineering at University of Wollongong. Currently, he is Distinguished Professor of Civil Engineering and Director of Transport Research Centre at the University of Technology Sydney (UTS).

Dr. Ana Heitor is a graduate from New University of Lisbon and obtained her master's and PhD degrees from Kyoto University (Japan) and the University of Wollongong (Australia), respectively. Having worked as a successful researcher and academic at the University of Wollongong for nearly a decade, Dr. Heitor is currently a lecturer at the University of Leeds, UK. Her research focuses on the behavior of compacted materials under unsaturated conditions. She is an associate editor of the *ICE Environmental Geotechnics and Soils and Foundations*. While she is still a young researcher, her research outcomes have been recognized by a number of national awards sponsored by the Australian Geomechanics and international awards, such as the 2017 John Carter Award and, more recently, the 2019 Winston Churchill Fellowship.

Associate Professor Jayan S. Vinod obtained his PhD from the Indian Institute of Science, Bangalore. He is currently an associate professor at the University of Wollongong, Australia. His research focuses on the numerical modeling of granular materials and soil dynamics, earthquake engineering and ground improvement techniques for rails and mine tailings. He is an associate editor of the *Journal of Materials in Civil Engineering* (ASCE) and one of the editors-in-chief of *International Journal of Geotechnical Earthquake Engineering* (IGI-Global: USA). He is a member of the Technical Committee (TC105: Geo-mechanics) of the International Society of Soil Mechanics and Geotechnical Engineering (ISSMGE). He has received the 2019 Australia-India (Early Mid-Career Fellowship) EMCR fellowship from the Australian Academy of Science. He has also delivered many keynote addresses and invited presentations and is actively involved in the Australian Research Council (ARC) and Cooperative Research Centres (CRC) on various research projects on ground improvement, railway geotechnics and mine tailings.

Chapter 1

Soil properties and classification

This chapter introduces different systems for classifying soils on the basis of their particle size distribution (PSD) or gradation and the plasticity of the fraction smaller than 425 µm.

1.1 Composition

Soils are particulate materials (Fig. 1.1), and their particles can be grouped in different sizes.

The building blocks of soil particles are typically composed the primary rock and clay minerals. The primary rock minerals are the products from mechanical weathering of the parent rock and thus share their chemical/mineral composition. They are typically angular or rounded and range from gravels to sand in size.

In contrast, the clay minerals are mainly products of chemical weathering, and their particles are very small (< 2 µm). They are composed of platelets that are the result of the combination of silica tetrahedron and octahedrons. Clays can be distinguished on the basis of the arrangement of the tetrahedral and octahedral sheets into layers. The way the structure is arranged can also give some indication of their likelihood to exhibit adverse behavior, e.g. montmorillonite clay experiences substantial changes in volume resulting from the water ingress in the interlayer spaces. In clay materials, the way the platelets are geometrically arranged, or fabric, also plays an important role in the expected behavior of the material. Figure 1.2 illustrates the main types of fabric observed for clays. For instance, while a clay prepared at relatively low water contents will tend to have a predominantly flocculated fabric (Fig. 1.2b), for larger water contents, the fabric will be predominately dispersed (Fig. 1.2a). Other types of common clay structures are illustrated in Figure 1.2c–e.

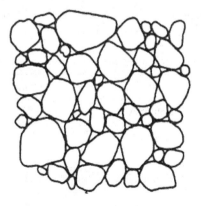

Figure 1.1 Example of grain structure (after Knappett and Craig, 2012, reproduced with permission from CRC Press).

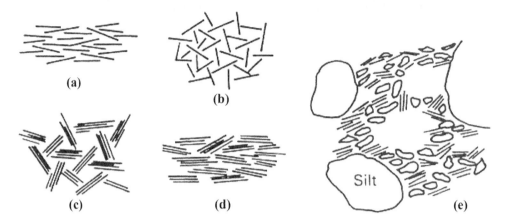

Figure 1.2 Clay structures: (a) dispersed; (b) flocculated; (c) bookhouse; (d) turbostratic; (e) example of a natural clay (after Knappett and Craig, 2012 reproduced with permission from CRC Press).

1.2 Classification

The purpose of the classification is to place a soil in one of the limited number of groups on the basis of its grading and plasticity. Since these characteristics are independent of the particular condition in which the soil occurs, they can assist in the diagnosis of potential adverse behavior of soil (e.g. excessive deformation, insufficient bearing capacity, erosion, piping, etc.) used as construction material.

Most systems of classification are based on the particle sizes that recognize three main classes (Table 1.1):

a. Oversize
b. Coarse grained
c. Fine grained

Oversize and coarse-grained soils are classified on the basis of the size and distribution. Particle-size distribution (gradation) is a descriptive term referring to the proportions by dry mass of a soil distributed over specified particle-size ranges that is typically determined via sieve analysis and/or sedimentation method. The sieve analysis is used for determining gradation of particles having a nominal size larger than the 74 μm sieve, whereas sedimentation or hydrometer method is used for the fraction finer than the 74 μm sieve and larger than

Table 1.1 Standard range of particle sizes proposed by AS 1726 (2017).

Fraction	Components	Subdivision	Size (mm)
Oversize	Boulders	-	>200
	Cobbles	-	63–200
Coarse-grained soil	Gravel	Course	19–63
		Medium	6.7–19
		Fine	2.36–6.7
	Sand	Course	0.6–2.36
		Medium	0.21–0.6
		Fine	0.075–0.21
Fine-grained soil	Silt	-	0.002–0.075
	Clay	-	<0.002

about 0.2 μm. The results are presented as the mass percent finer versus the logarithm of the particle diameter. Alternatively, the gradation of the fraction finer than 74 μm can also be determined using laser diffraction particle analysis. However, due care must be exercised as this method computes the percentage of particles by volume rather than by mass.

The important features of a particle size distribution may be expressed in terms of the uniformity coefficient (C_u) and the coefficient of curvature (C_c). Three sizes may be determined from the PSD curve, D_{10}, D_{30} and D_{60}, and they represent the size such that 10%, 30% and 60% (by weight) of the sample consists of particles having a smaller nominal diameter, respectively. The D_{10} is also termed effective size and can be used to estimate the permeability of a soil.

$$C_u = \frac{D_{60}}{D_{10}} \tag{1.1}$$

$$C_c = \frac{\left(D_{30}\right)^2}{D_{60} \times D_{10}} \tag{1.2}$$

The fine-grained soils are classified on the basis of their plasticity. This can be determined through consistency or Atterberg limits. Three limits distinguish the changes in state observed for a soil when there is a variation in water content, i.e.:

a. the *liquid limit* (*LL*) is the water content at which the soil passes from plastic to liquid state,
b. the *plastic limit* (*PL*) is the lowest water content at which the soil remains in plastic state and
c. the *shrinkage limit* (*SL*) is the water content at which further loss of water will not cause further reduction in the volume of the soil.

A number of indices may be derived from these limits, e.g. plasticity index (*PI*), liquidity index (*LI*) and soil activity (*A*).

$$PI = LL - PL \tag{1.3}$$

$$LI = \frac{w - PL}{PI} \tag{1.4}$$

$$A = \frac{PI}{\% \text{ weight of clay particles } (<2\,\mu m)} \tag{1.5}$$

1.3 Unified soil classification system

In this classification, the soils are designated by two letters, in which the first letter represents the main soil type and second denotes a qualifying subdivision (Table 1.2). Soils are divided by the percentage finer than 75 μm. For soils having more than 50% retained in the 75 μm sieve, the prefix letters allotted are G = Gravel, S = sand. The suffix letters are allotted on the basis of C_u and C_c values as follows:

$C_u > 4$ and $1 < C_c < 3$ for gravels
$C_u > 6$ and $1 < C_c < 3$ for sand

If the soil meets the above requirements, the suffix letter is W = well graded, else it is P = poorly graded (Table 1.2). For sands having more than 12% of fines, the suffix letters are either C = Clay or M = silt. For soils having more than 50% passing in the 75 μm, the prefix letters allotted are C = clay (inorganic), M = silt, O = organic clay and Pt = Peat (Table 1.3).

The suffix letters are allotted based on *LL* and *PI* using the Casagrande plasticity chart (Fig. 1.3), where L = Low plasticity, I = Intermediate and H = High plasticity.

Table 1.2 Classification of coarse-grained soils (modified after AS 1726, 2017).

Major divisions	Group symbol	Typical names	Field identification sand and gravels	Laboratory classification**	
Coarse-grained soil (>65% of material is greater than 0.075 mm)*					
GRAVELS (>50% coarse fraction is larger than 2.36 mm)	GW	Gravels and gravel–sand mixtures, little or no fines	Wide range in grain size and substantial amounts of all intermediate sizes, not enough fines to bind coarse grains, no dry strength	≤5% fines	$C_u > 4$ $1 < C_c < 3$
	GP	Gravels and gravel–sand mixtures, little or no fines, uniform gravels	Predominantly one size or range of sizes with some intermediate sizes missing, not enough fines to bind coarse grains, no dry strength	≤5% fines	Fails to comply with above
	GM	Gravel–silt mixtures and gravel–sand–silt mixtures	Materials with excess of nonplastic fines, zero to medium dry strength	≥12% fines, fines are silty	–
	GC	Gravel–clay mixtures and gravel–sand–clay mixtures	Materials with excess of plastic fines, medium to high dry strength	≥12% fines, fines are clayey	
SANDS (>50% coarse fraction is smaller than 2.36 mm)	SW	Sands and gravel–sand mixtures, little or no fines	Wide range in grain size and substantial amounts of all intermediate sizes, not enough fines to bind coarse grains, no dry strength	≤5% fines	$C_u > 6$ $1 < C_c < 3$
	SP	Sands and gravel–sand mixtures, little or no fines	Predominantly one size or range of sizes with some intermediate sizes missing, not enough fines to bind coarse grains, no dry strength	≤5% fines	Fails to comply with above
	SM	Sand–silt mixtures	Materials with excess of nonplastic fines, zero to medium dry strength	≥12% fines, fines are silty	–
	SC	Sand–clay mixtures	Materials with excess of plastic fines, medium to high dry strength	≥12% fines, fines are clayey	

* Excludes oversize fraction.

** For fines content between 5% and 12%, the soil has dual classification having two group symbols separated by a dash, e.g. for a sand having 5%–12% clay fines, then the classification is SP-SC. Soils dominated by boulders, cobles or peat (Pt) are described separately.

Table 1.3 Classification of fine-grained soils (modified after AS 1726, 2017).

Major divisions	Group symbol	Typical names	Field identification of silts and clays			Laboratory classification**
			Dry strength	Dilatancy	Toughness	% <0.075 mm
Fine-grained soils (>35% of material is less than 0.075 mm*)						
SILTS and **CLAYS** (low to medium plasticity)	ML	Inorganic silts and very fine sands, rock flour, silty or clayey fine sands or silts with low plasticity	None to low	Slow to rapid	Low	Below A line
	CL, CI	Inorganic clays of low to medium plasticity, gravelly clays, sandy clays	Medium to high	None to slow	Medium	Above A line
SILTS and **CLAYS** (high plasticity)	OL	Organic silts	Low to medium	Slow	Low	Below A line
	MH	Inorganic silts	Low to medium	None to slow	Low to medium	Below A line
	CH	Inorganic clays of high plasticity	High to very high	None	High	Above A line
	OH	Organic clays of medium to high plasticity, organic silts	Medium to high	None to very slow	Low to medium	Below A line
Highly organic soils	Pt	Peat, highly organic soils	–	–	–	–

* Excludes oversize fraction.
** See Figure 1.3.

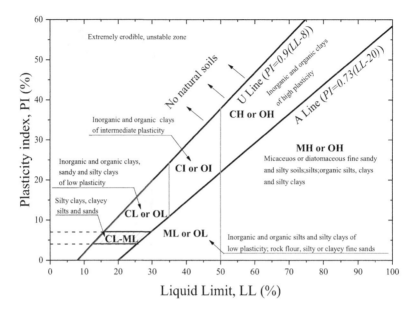

Figure 1.3 Modified Casagrande's plasticity chart showing the classification based on soil plasticity proposed by AS 1726 (2017).

Table 1.4 Results of the sieve analysis.

Sieve size	Mass retained (g)
3.35 mm	0
2.00 mm	2.8
1.18 mm	13.8
600 µm	53.6
425 µm	66
300 µm	31.5
212 µm	21.3
150 µm	11.9
75 µm	10.7

Table 1.5 Results of the sieve analysis – solution.

Sieve size	Mass retained (g)	Mass retained (% of total weight)	Cumulative mass percentage
3.35 mm	0	0.00	100.00
2.00 mm	2.8	1.31	98.69
1.18 mm	13.8	6.44	92.25
600 µm	53.6	25.02	67.23
425 µm	66	30.81	36.41
300 µm	31.5	14.71	21.71
212 µm	21.3	9.94	11.76
150 µm	11.9	5.56	6.21
75 µm	10.7	5.00	1.21
Pan	2.6	1.21	0.00

Worked-out example 1.1: A soil sample was collected in a trial pit to determine the soil particle size distribution. The dry mass retained in a given sieve size aperture was recorded and is listed in Table 1.4. Plot the PSD curve, knowing that the total mass of sample was 214.2 g. The mass passing through the 75 microns sieve was 2.6 g. Determine the coefficient of uniformity and curvature.

Solution: Before plotting the PSD, the equivalent percentages of the differences in size ranges need to be calculated.

To determine the mass retained as a percentage of the total mass, the ratio between the different mass weights retained in the different size ranges by total mass needs to be computed,

e.g. mass retained % (2–3.35 mm) = 2.8 / 214.2 = 1.31%

Recalculating for the other size ranges, the results shown in Table 1.5 are obtained. For each of the size ranges, percentage of mass retained can be converted to cumulative mass, as shown in Table 1.5. The cumulative mass percentage or percent finer can then be plotted on the y-axis, whereas the corresponding size is plotted on the x-axis (logarithmic scale) as shown in Figure 1.4.

To determine the coefficient of uniformity and curvature, the diameter of the particles corresponding to 10, 30 and 60% of the mass retained need to be evaluated in the manner shown in Figure 1.4.

$$D_{10} = 0.2 \text{ mm}, D_{30} = 0.38 \text{ mm and } D_{60} = 0.54 \text{ mm}$$

Then

$$C_u = \frac{D_{60}}{D_{10}} = 2.7 \qquad C_c = \frac{\left(D_{30}\right)^2}{D_{60} \times D_{10}} = 1.34$$

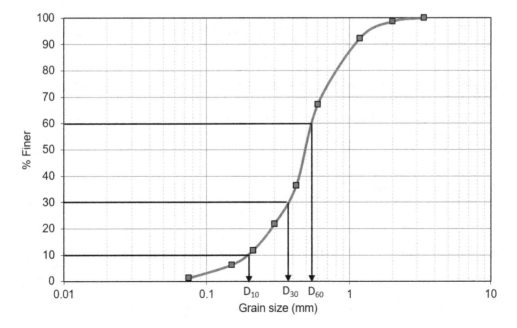

Figure 1.4 Plot of the PSD analysis.

Worked-out example 1.2: From the following gradation characteristics (Table 1.6), sketch the grading curves for the three soils, and identify the soil type (e.g. uniform medium sand). Discuss how you would identify these soils in the field by sight or touch.

Table 1.6 Gradation characteristics.

Soil	D_{10} (mm)	C_u	C_c
A	0.28	1.50	0.87
B	0.088	19.9	0.80
C	0.009	167	0.12

Solution: For sketching the grading curve, we require more information about the material, given that both C_u and C_c values are known, then D_{30} and D_{60} may be calculated as follows:

$$C_c = \frac{(D_{30})^2}{D_{60} \times D_{10}} = 1.25$$

Soil A: $C_u = \dfrac{D_{60}}{D_{10}}$ with $1.5 = \dfrac{D_{60}}{0.28}$ then $D_{60} = 0.42$

similarly $C_c = \dfrac{(D_{30})^2}{D_{60} \times D_{10}}$ with $0.87 = \dfrac{(D_{30})^2}{0.42 \times 0.28}$ then $D_{30} = 0.319$

Recalculating for Soil B and C in a similar manner, the following results are obtained (Table 1.7).

The grading curve is represented in Figure 1.5. Soil A is a uniform fine sand, Soil B is a nonuniform sand.

Table 1.7 D_{30} and D_{60} results.

Soil	D_{10} (mm)	C_u	C_c	D_{30} (mm)	D_{60} (mm)
A	0.28	1.50	0.87	0.32	0.42
B	0.088	19.9	0.80	0.35	1.75
C	0.009	167	0.12	0.04	1.50

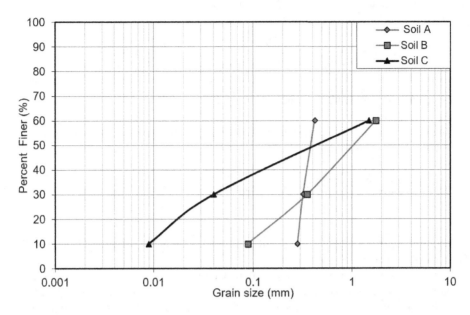

Figure 1.5 PSD curve based on D_{10}, D_{30} and D_{60} sizes.

Worked-out example 1.3: Activity of clay is useful in identifying the type of clay minerals as given in Table 1.8.

The water content of a soil sample is 45%, its liquid limit is 70% and plastic limit is 35%. It contains approximately 80% by weight of clay.

i.   What type of clay is this sample?
ii.   Explain the significance of its liquidity index.

Table 1.8 List of type of clay minerals and associated activity.

Clay mineral	Activity
Montmorillonite	> 5
Illite	$\cong 0.9$
Kaolinite	$\cong 0.4$

Solution:
i.   To identify the likely type of the clay in terms of its activity, we need to compute the activity of clay based on its plasticity index (*PI*) and %weight of clay particles.

Thus, for a soil having $LL = 70$ and $PL = 35$, then $PI = LL - PL = 70-35 = 35\%$, then

$$\text{Soil Activity, } A = \frac{PI}{\% \text{ weight of clay particles} \left(< 2\mu m \right)} = \frac{0.35}{0.8} = 0.44$$

which indicates it is most likely composed of kaolinite.

ii.   The liquidity index can be determined based on the consistency limits and water content of the sample, as follows:

$$LI = \frac{w - PL}{PI} = \frac{0.45 - 0.35}{0.35} = 0.29 \text{ or } 29\%$$

The liquidity index is a measure that enables the evaluation of the current moisture condition of a soil with respect to its index limits.

A cohesive soil with a natural water content of the same order as its liquid limit (LI approaching 1) will be a very soft material, while with a natural water content of the same order as its plastic limit (LI approaching 0 or even negative) will be a stiffer material. This can have substantial impact in the way the soil behaves as a construction material (e.g. bearing capacity for construction plant).

Worked-out example 1.4: A cohesive material was tested for liquid and plastic limits using standard methods. The results obtained using the cone penetrometer test apparatus are shown in Table 1.9, e.g. the water content and associated cone penetrations values. The plastic limit was 14.5%.

Describe and classify the soil.

Table 1.9 Results of a cone liquid limit test.

Water content (%)	23.5	24.5	25.1	26.0	26.8
Cone penetration (mm)	12.0	15.6	17.7	21.0	25.6

Solution: The cone liquid limit is the water content for which 20 mm penetration is observed. Typically, it is difficult to determine the water content for which penetration is exactly 20 mm. Instead, a series of water content levels are tested and their penetration recorded. The results presented in Table 1.9 can be represented in a graph, e.g. on the x-axis, the water content results are plotted for the corresponding cone penetration on the y-axis (Fig. 1.6).

Thus the liquid limit is approximately 25.6% and $PI = 25.6-14.5 = 11.1\%$; the soil can be classified as low-plasticity clay or CL based on the plasticity chart shown in Figure 1.7.

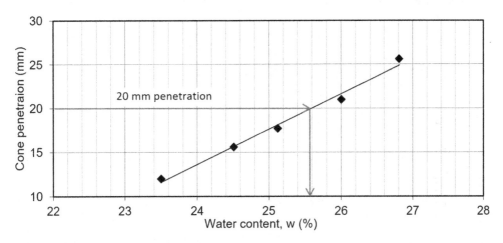

Figure 1.6 Graphical representation of the cone penetrometer test.

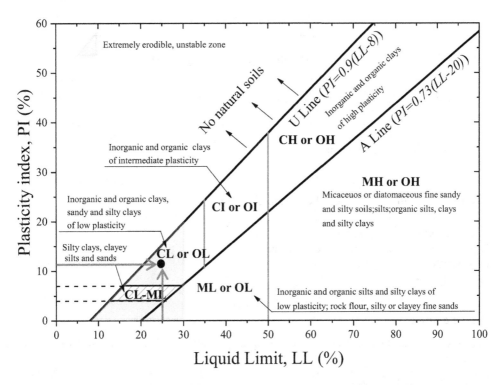

Figure 1.7 Modified Casagrande's plasticity chart showing the classification based on soil plasticity proposed by AS 1726 (2017).

Worked-out example 1.5: Based on the particle size distributions displayed in Figure 1.8, classify the soils A to D.

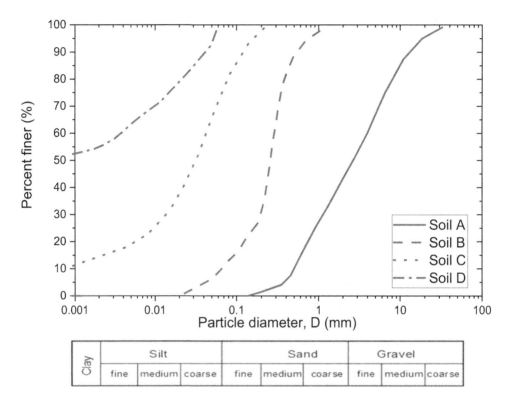

Figure 1.8 Particle size distribution.

Solution: The different soils can be classified as follows:

A → Well-graded sandy gravel
B → Uniform (poorly graded) medium sand
C → Sandy silt
D → Silty clay

References

Knappett, J. & Craig, R.F. (2012) *Craig's Soil Mechanics*, 8th edition. CRC Press, ISBN:9780415561266, p. 584.

Standards Australia (2017) *Geotechnical Site Investigations*. Australian Standard AS1726-2017, Standards Australia, Sydney.

Chapter 2

Weight–volume relations and compaction

Soil is a particulate material that encloses voids or pores. It can be found in three different states, dry (voids filled with air), saturated (voids filled with water) or partially saturated (voids filled with air and water) as illustrated in Figure 2.1 and Figure 2.2.

2.1 Weight and volume relations

a. Void ratio (e) represents the ratio between the volume of the voids (V_v) and the volume occupied by the solid particles (V_s) or

$$e = \frac{V_v}{V_s} \tag{2.1}$$

b. Porosity represents the ratio between the volume of the voids and the total volume (V) or

$$n = \frac{V_v}{V} \text{ and } n = \frac{e}{e+1} \text{ or } e = \frac{n}{1-n} \tag{2.2}$$

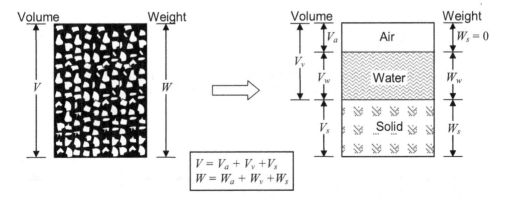

$$V = V_a + V_v + V_s$$
$$W = W_a + W_v + W_s$$

Figure 2.1 General weight–volume relations for a partially saturated soil element.

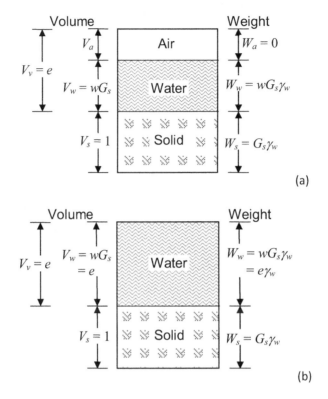

Figure 2.2 (a) Partially and (b) fully saturated soil elements (V_s = 1).

c. Degree of saturation (S_r) represents the ratio between the volume of water (V_w) and the volume of the voids or

$$S_r = \frac{V_w}{V_v} \qquad (2.3)$$

d. Water content (w) represents the ratio between the weight (W_w) or mass (M_w) of the water in the soil and the weight (W_s) or mass of the solid particles (M_s) or

$$w = \frac{W_w}{W_s} \quad \text{or} \quad w = \frac{M_w}{M_s} \qquad (2.4)$$

e. Air voids content (A_v) represents the ratio between the volume of air (V_a) and total volume or

$$A_v = \frac{V_a}{V} = \frac{e(1 - S_r)}{1 + e} \qquad (2.5)$$

f. Bulk unit weight (γ_b) or density (ρ_b) represents the ratio between the total weight (W) or mass (M) and the total volume or

$$\gamma_b = \frac{W}{V} \text{ or } \rho_b = \frac{M}{V} \text{ and } \gamma_b = \rho_b g \tag{2.6}$$

(g = gravitational acceleration constant, e.g. 9.81 m/s^2)

g. Dry unit weight (γ_d) or density (ρ_d) represents the ratio between the weight (W_s) or mass (M_s) of the solid particles and the total volume or

$$\gamma_d = \frac{W_s}{V} \text{ or } \rho_d = \frac{M_s}{V} \text{ and } \gamma_d = \rho_d g \tag{2.7}$$

h. Specific gravity of the solid particles (G_s) represents the ratio of the mass of a given volume of solid particles to the mass of water for the same volume or

$$G_s = \frac{M_s}{V_s}\frac{1}{\rho_w} = \frac{W_s}{V_s}\frac{1}{\gamma_w} \tag{2.8}$$

where $\rho_w = 1\,\text{Mg/m}^3$ and $\gamma_w = 1\times 9.81 = 9.81\,\text{kN/m}^3$ represent the density and unit weight of water, respectively.

i. Specific volume (v) is $v = 1 + e$
j. Common relations using the weight–volume indices:

$$S_r e = w G_s \tag{2.9}$$

$$\gamma_b = \gamma_d (1 + w) \text{ or } \rho_b = \rho_d (1 + w) \tag{2.10}$$

$$\gamma_b = \frac{W}{V} = \frac{G_S + S_r e}{1+e}\gamma_w \tag{2.11}$$

$$\gamma_{sat} = \frac{M}{V} = \frac{G_S + e}{1+e}\gamma_w \tag{2.12}$$

$$\gamma_d = \frac{G_s \gamma_w}{1+e} \text{ or } \rho_d = \frac{G_s \rho_w}{1+e} \tag{2.13}$$

2.2 State of compaction

During construction, soil is usually compacted to meet a specified laboratory criterion based on the optimum values, typically established according to the dry density–water content relationship, for a given compaction energy (i.e. standard or modified compaction, Table 2.1).

The maximum dry density ($\gamma_{d,max}$) is achieved at the optimum moisture content (OMC) (Fig. 2.3). For a given compaction energy level, the dry unit weight increases as the moisture content increases to the OMC. Beyond this point (i.e. wet side of the compaction plane), the dry unit weight decreases with increasing water content (w). This tendency can be explained considering the interactions between water, air and solid phases.

Table 2.1 Differences between Proctor standard and modified compaction tests.

	Standard (ASTM D698) (AS 1289.5.1.1–2003)	Modified (ASTM D1557)
Hammer weight, N (lb)	24.5 (5.5)	44.5 (10)
Height of hammer fall, mm (in.)	305 (12)	457 (18)
Number of layers	3	5
No. of blows/layer*	25	25
Mold volume,* m³ (ft³)	0.000944 (1/30)	
Soil	(-) No. 4 sieve	
Compaction energy (CE), kJ/m³ (lbft/ft³)	595 (12400)	2698 (56250)

* Using the 102-mm- (4-in.) diameter mold

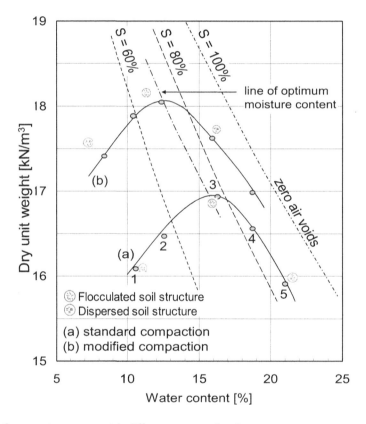

Figure 2.3 Compaction curve with different energy levels.

On the dry side of the compaction plane (points located below the line of optima), the suction that acts on the particle contacts to oppose slippage is high, and the compaction process yields low dry unit weights and a flocculated structure. The progressive addition of moisture reduces suction and facilitates particle slippage. Thus, the soil experiences higher dry unit weights until it reaches its maximum at OMC, represented by the line of optimum moisture content in Figure 2.3. Beyond the OMC, the air phase becomes discontinuous, i.e. air is occluded in bubbles. In this condition, any applied external compaction energy is likely to be supported by the water phase in the soil, as compaction occurs over a relatively short period and the system is undrained.

2.2.1 Relative compaction

Relative compaction is a method of specifying the end results that are to be achieved in the field, where it is also known as end-product specification. The minimum relative compaction is governed by strategic importance of the fill and the desired project objectives (AS 3798; Standards Australia 2007) and requirements, between 90% and 95% relative compaction is typically considered acceptable.

$$\text{Relative compaction (\%)} = \frac{\gamma_{d,field}}{\gamma_{d,\max}} \times 100 = \frac{\rho_{d,field}}{\rho_{d,\max}} \times 100 \qquad (2.14)$$

For instance, in Figure 2.4, a minimum of 95% relative compaction is adopted to define the field water content working range. However, often the water content variation is also specified in relation to the OMC, i.e. 2% ± OMC (e.g. AS 3798; Standards Australia 2007).

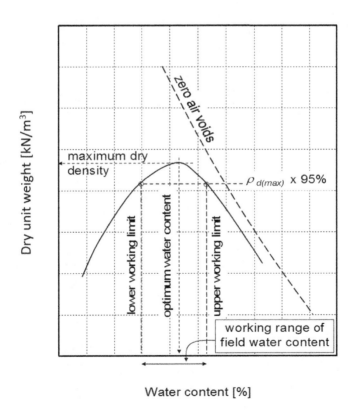

Figure 2.4 Compaction curve and definition of the field working range based on the relative compaction.

2.2.2 Effect of soil type on the compaction curve

Typically, well-graded granular soils (e.g. GW) attain the highest γ_d and lowest OMC, whereas silt (ML) and organic clays (CH) attain the lowest γ_d and highest OMC (Fig. 2.5).

2.2.3 Alternative compaction curve representation for waste materials

The compaction data is conventionally represented in terms of $\gamma_d - w$, which is directly computed from the laboratory standard Proctor compaction tests (e.g. AS 1289.5.1.1; Standards Australia 2017). The main advantage of using these two variables (i.e. $\gamma_d - w$)

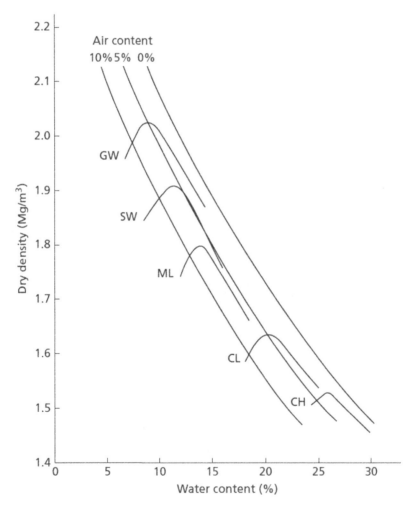

Figure 2.5 Compaction curves for different soils (after Knappett and Craig, 2012 reproduced with permission from CRC Press).

lies in their simple determination based on the experimental results, that is to say, wet unit weight or bulk unit weight (γ) can be simply calculated considering the mass over volume ratio of the soil in the mold and dry unit weight can be easily obtained considering the water content (Eq. 2.15).

$$\gamma_d = \frac{\gamma}{(1+w)} \tag{2.15}$$

However, for mixtures involving materials having different G_s values (for instance, granular waste materials), the compaction data represented in the conventional $\gamma_d - w$ reflects the weight percentage of the different materials rather than the compaction efficiency or degree of densification.

Furthermore, for different blend ratios of materials having different G_s values, several zero air voids (ZAV) lines (Eq. 2.16) need to be plotted, and the representation of additional saturation lines is cumbersome.

$$\gamma_d = \frac{G_s \gamma_w}{1 + \dfrac{wG_s}{S_r}} \tag{2.16}$$

where γ_w is the unit weight of the water and S_r is the degree of saturation ($S_r = 1$ for fully saturated conditions).

To avoid these limitations, the compaction data expressed in terms $\gamma_d - w$ may be represented in terms of the equivalent variables of void ratio (e) and water ratio (e_w) determined based on the value of G_s, as follows:

$$e = \frac{\gamma_w G_s}{\gamma_d} - 1 \tag{2.17}$$

$$e_w = eS_r = wG_s \tag{2.18}$$

The use of alternative representation for waste granular materials has significant advantages in relation to the conventional $\gamma_d - w$ representation (Fig. 2.6), since G_s is incorporated in the determination of both e and e_w.

Furthermore, in this representation, the different degree of saturation lines concerning different specific gravity values (Fig. 2.6) are now unique lines (i.e. independent of the value of G_s) having a gradient of $1/S_r$ (Eq. 2.19). This can be advantageous in locating the compaction states for the different blend ratios that prescribe to a given range of percentage of air voids (A_v), for instance, shown in Eq. 2.20.

$$e = \frac{1}{S_r} e_w \tag{2.19}$$

$$A_v = n(1 - S_r) \tag{2.20}$$

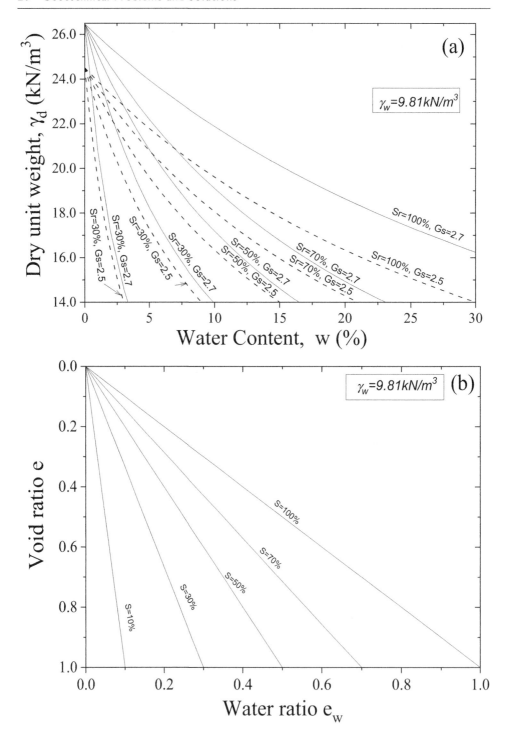

Figure 2.6 Degree of saturation lines across the compaction plane in terms of (a) $\gamma_d - w$ and (b) $e-e_w$.

Worked-out example 2.1: Derive the following weight–volume relations:

a. $n = \dfrac{e}{e+1}$

b. $S_r e = wG_s$

c. $\gamma_b = \gamma_d \left(1 + w\right)$ or $\rho_b = \rho_d \left(1 + w\right)$

d. $\gamma_b = \dfrac{W}{V} = \dfrac{G_S + S_r e}{1+e} \gamma_w$

e. $\gamma_{sat} = \dfrac{M}{V} = \dfrac{G_S + e}{1+e} \gamma_w$

f. $\gamma_d = \dfrac{G_s \gamma_w}{1+e}$ or $\rho_d = \dfrac{G_s \rho_w}{1+e}$

Solution:

a. Expanding the void ratio terms and multiplying the denominator by $\dfrac{V_s}{V_s}$ after simplification, we obtain the porosity relationship:

$$n = \frac{e}{e+1} = \frac{\dfrac{V_v}{V_s}}{\dfrac{V_v}{V_s} + \dfrac{V_s}{V_s}} = \frac{\dfrac{V_v}{V_s}}{\dfrac{V_v + V_s}{V_s}} = \frac{V_v}{V_v + V_s} = \frac{V_v}{V} = n$$

b. We can start in either side. Expanding the S_r and e terms and rearranging, we have:

$$S_r \times e = \frac{V_w}{V_v} \times \frac{V_v}{V_s} = \frac{V_w}{V_s} \times \frac{V_v}{V_v} = \frac{V_w}{V_s}$$

Including $\dfrac{W_w}{W_w} \times \dfrac{W_s}{W_s}$ and then rearranging, we can easily obtain the expanded forms of the product of specify gravity and water content:

$$S_r \times e = \frac{V_w}{V_s} \times \frac{W_w}{W_w} \times \frac{W_s}{W_s} = \frac{W_s}{V_s} \times \frac{V_w}{W_w} \times \frac{W_w}{W_s} = \frac{\gamma_s}{\gamma_w} \times \frac{W_w}{W_s} = G_S w$$

c. Expanding the bulk unit weight in terms of the weight components (water and solid particles):

$$\gamma_b = \frac{W}{V} = \frac{W_s}{V} + \frac{W_w}{V}$$

Including $\dfrac{W_s}{W_s}$ in the second term and then rearranging:

$$\gamma_b = \frac{W_s}{V} + \frac{W_w}{V} \frac{W_s}{W_s} = \frac{W_s}{V} + \frac{W_w}{W_s} \frac{W_s}{V}$$

Selecting the common fraction $\dfrac{W_s}{V}$ and rearranging, we can easily obtain the expanded form of the dry unit weight and water content as follows:

$$\gamma_b = \frac{W_s}{V}\left(1+\frac{W_w}{W_s}\right) = \gamma_d\left(1+w\right)$$

d.   Expanding the bulk unit weight term and rearranging, we obtain:

$$\gamma_b = \frac{W}{V} = \frac{W_s + W_w}{V_s + V_v} = \frac{W_s\left(1+\dfrac{W_w}{W_s}\right)}{V_s\left(1+\dfrac{V_v}{V_s}\right)}$$

Noticing that $\gamma_s = \dfrac{W_s}{V_s}$ and that $G_s = \dfrac{\gamma_s}{\gamma_w}$ or $\gamma_s = G_s\gamma_w$ and recognizing $w = \dfrac{W_w}{W_s}$ and $e = \dfrac{V_v}{V_s}$, then we can obtain:

$$\gamma_b = G_s \times \gamma_w\left(\frac{1+w}{1+e}\right) = \left(\frac{G_s + G_s w}{1+e}\right)\gamma_w$$

Using the expression derived earlier $S_r e = wG_s$ then, we can obtain the following relationship:

$$\gamma_b = \left(\frac{G_s + S_r e}{1+e}\right)\gamma_w$$

e.   From the derivation in d, we have the relationship for unsaturated soil, as follows:

$$\gamma_b = \left(\frac{G_s + S_r e}{1+e}\right)\gamma_w,$$ to calculate saturated unit weight, in which all pores are filled with water $S_r = 1$, then the expression can be simplified as follows:

$$\gamma_{sat} = \frac{G_S + (1\times e)}{1+e}\gamma_W = \frac{G_S + e}{1+e}\gamma_W$$

f.   Only the derivation for dry unit weight is demonstrated. Expanding the dry unit weight term, we obtain:

$$\gamma_d = \frac{W_s}{V}$$

Knowing that $1 + e = 1 + \dfrac{V_v}{V_s} = \dfrac{V_s}{V_s} + \dfrac{V_v}{V_s} = \dfrac{V}{V_s}$ thus $V = (1+e)V_s$, then

$$\gamma_d = \frac{W_s}{(1+e)V_s}$$

Noticing that $\gamma_s = \dfrac{W_s}{V_s}$ and that $G_s = \dfrac{\gamma_s}{\gamma_w}$ or $\gamma_s = G_s \gamma_w$, we can obtain:

$$\gamma_d = \dfrac{G_s \gamma_w}{1+e}$$

Worked-out example 2.2: The following index properties were determined for two soils, A and B, and are listed in Table 2.2. Which of these soils (a) has a greater void ratio? (b) has the greater dry density? (c) has the greater bulk or wet density? Justify your answer.

Table 2.2 Index properties of soil A and B.

Soil	A	B
Water content (%)	25.0	15.0
Specific gravity	2.67	2.7
Degree of saturation	1.0	0.85

Solution:

a. $w = \dfrac{W_w}{W_s} = \dfrac{V_w \rho_w}{V_s G_s \rho_w} \dfrac{V_v}{V_v} = \dfrac{S_r e}{G_s}$ then

Soil A: $0.25 = \dfrac{1 \times e}{2.67}$ and $e = 0.67$

Soil B: $0.15 = \dfrac{0.85 \times e}{2.7}$ and $e = 0.48$

Hence, Soil A has the greatest void ratio.

b. $\rho_d = \dfrac{G_s \rho_w}{1+e}$ and $\rho_w = 1\ \mathrm{Mg/m^3}$ then

Soil A: $\rho_d = \dfrac{2.67 \times 1.0}{1+0.67}$ and $\rho_d = 1.6\ \mathrm{Mg/m^3}$

Soil B: $\rho_d = \dfrac{2.7 \times 1.0}{1+0.48}$ and $\rho_d = 1.82\ \mathrm{Mg/m^3}$

Hence, Soil B has the greatest dry density.

c. $\rho_b = \rho_d (1+w)$, then

Soil A: $\rho_b = 1.6\,(1+0.25)$ and $\rho_b = 2.0\ \mathrm{Mg/m^3}$

Soil B: $\rho_b = 1.82\,(1+0.15)$ and $\rho_b = 2.1\ \mathrm{Mg/m^3}$

Hence, Soil B has the greatest bulk and wet density but is very similar to Soil A.

Worked-out example 2.3: A fully saturated soil sample, collected at a field site, has a volume of 185 ml and mass of 331 g. The specific gravity of the solid particles is 2.7. Determine the void ratio, porosity, water content and bulk density.

Solution: As the sample is fully saturated, all the voids are occupied by water, and degree of saturation is 1. The bulk density can be easily determined:

$$\rho_b = \frac{M}{V} \text{ then } \rho_b = \frac{331}{185} = 1.79 \, \text{g/cm}^3$$

We can derive the relationship for expressing bulk unit weight in terms of void ratio and specific gravity of the soil particles, considering $\gamma_s = \frac{W_s}{V_s}$ and that $G_s = \frac{\gamma_s}{\gamma_w}$ or $\gamma_s = G_s \gamma_w$ and $S_r e = wG_s$ and $S_r = 1$:

$$\rho_b = \frac{M}{V} = \frac{M_s + M_w}{V_s + V_v} = \frac{M_s \left(1 + \dfrac{M_w}{M_s}\right)}{V_s \left(1 + \dfrac{V_v}{V_s}\right)} = \frac{G_s \rho_w (1 + w)}{1 + e}$$

$$= \frac{G_s \rho_w + G_s w \rho_w}{1 + e} = \frac{G_s \rho_w + eS_r \rho_w}{1 + e} = \rho_w \frac{G_s + e}{1 + e}$$

Then,

$$1.79 = 1 \times \frac{2.7 + e}{1 + e} \text{ and } e = 1.15 \text{ and porosity can be easily determined using } n = \frac{e}{e + 1}$$

$$n = 1 \times \frac{1.15}{1.15 + 1} \text{ and } n = 0.53 \text{ or } n = 53\%$$

For determining water content, we can use the relation derived earlier:

$$\rho_b = \frac{G_s \rho_w (1 + w)}{1 + e}$$

we can then obtain

$$1.79 = \frac{2.7 \times 1 \times (1 + w)}{1 + 1.15} \text{ and } w = 0.43 \text{ or } = 43\%$$

Worked-out example 2.4: A soil deposit to be used for construction of an embankment has an average dry density of 1602 kg/m³ and a moisture content of 10%. The compacted soil in the embankment is to have a dry density of 1842 kg/m³. During the process of compaction, no water is to be added to the soil.

Determine the volume of soil to be excavated from the natural soil deposit for 76,455 m³ of the completed embankment and the wet mass of the soil to be excavated. Assume that no soil moisture is lost during excavation, handling and compaction.

Solution: The volume of material in loose state can be determined using the ratio between the compacted dry density and loose dry density (consider M_s is constant, as no mass is lost during compaction), as follows:

$$\frac{\rho_{d,compacted}}{\rho_{d,loose}} = \frac{M_s}{V_{compacted}} \times \frac{V_{loose}}{M_s} \text{ then } V_{loose} = \frac{1842}{1602} \times 76455 = 87909 \, \text{m}^3$$

Hence, the volume of soil to be excavated is 87,909 m³.

To determine the wet mass, we need to calculate the bulk density first, using the relation between bulk and dry density as follows:

$$\rho_b = \rho_d \ (1+w) \text{ then } \rho_b = 1602 \times (1+0.1) = 1762.2 \text{ kg/m}^3$$

Then we can calculate wet mass considering the loose volume determined earlier.

$$\rho_b = \frac{M}{V} \text{ then } M = \rho_b V$$

$$M = 1762.2 \ \times 87909 = 154913 \text{ tons}$$

Hence the wet mass required is 154,913 tons.

Worked-out example 2.5: The density of a rolled earth fill was checked in the field by excavating a hole in it and carefully removing the excavated material, which had a mass of 8.47 kg and, after oven drying, 6.80 kg. The volume of the hole was determined by measuring the mass of loosely deposited sand required to fill it, and this was 6.2 kg. The bulk density of similarly deposited sand was found to be 1475 kg/m³. The specific gravity of solid constituents was 2.74. What is the water content, dry density and degree of saturation of the fill?

Solution: The water content can be determined considering the humid and dry mass of the excavated material as follows:

$$w = \frac{M_w}{M_s} = \frac{(8.47 - 6.8)}{6.8} = 0.246 \text{ or } 24.6\%$$

To determine the dry density, we must first determine the bulk density of the excavated earth fill. The bulk density of sand used to fill the hole is 1475 kg/m³, and the associated mass is 6.2 kg; thus the volume of the hole can be computed as follows:

$$\rho_b = \frac{M}{V} \Leftrightarrow 1475 = \frac{6.2}{V} \Leftrightarrow V = 4.2 \times 10^{-3} \text{ m}^3$$

Knowing the mass of soil that was excavated (8.47 kg), bulk density of the excavated material can be determined as follows:

$$\rho_b = \frac{M}{V} = \frac{8.47}{4.2 \times 10^{-3}} = 2015 \text{ kg/m}^3$$

Using the following relationship, dry density can then be determined:

$$\rho_b = \rho_d \ (1+w)$$

Rearranging we obtain:

$$\rho_d = \frac{\rho_b}{(1+w)} = \frac{2015}{(1+0.246)} = 1617.2 \text{ kg/m}^3$$

Using the following relationship, we can determine first the void ratio (note density of the water, $\rho_w = 1000\,\text{kg/m}^3$):

$$\rho_d = \frac{G_s \rho_w}{1+e} \Leftrightarrow 1617.2 = \frac{2.74 \times 1000}{1+e} \Leftrightarrow e = 0.69$$

Then the degree of saturation can be computed based on the following weight–volume relationship:

$$S_r e = w G_s \Leftrightarrow S_r = \frac{w G_s}{e} \Leftrightarrow S_r = \frac{0.246 \times 2.74}{0.69} = 0.977 \text{ or } 97.7\%$$

The water content of the fill is 24.6%, the dry density is 1617.2 kg/m³ and the degree of saturation is 97.7%.

Worked-out example 2.6: The following data is obtained in the determination of field (compaction) density.

a.   Weight of wet fill removed from test hole = 1942 g
b.   Weight of sand used to fill the hole and cone = 2744 g
c.   Density of sand = 1.60 g/cm³
d.   Weight of sand to fill cone = 1289.7 g (by calibration)

Determine the in-situ wet and dry unit weights of the fill if the entire sample is oven-dried and has a weight of 1708.7 g.

Solution: The water content can be first determined considering the wet and dry weights of the excavated material as follows:

$$w = \frac{M_w}{M_s} = \frac{(1942 - 1708.7)}{1708.7} = 0.137 \text{ or } 13.7\%$$

The density of sand used to fill the hole is 1.6 g/cm³, and the associated mass of sand stored in the hole is obtained by subtracting the weight of sand used to fill the hole and cone and the weight of sand to fill the cone
   (2744 − 1289.7 = 1454.3 g); thus the volume of the hole can be computed as follows:

$$\rho_b = \frac{M}{V} \Leftrightarrow 1.6 = \frac{1454.3}{V} \Leftrightarrow V = 908.94 \text{ cm}^3$$

Knowing the mass of soil that was excavated (1942 g), bulk density of the excavated material can be determined as follows:

$$\rho_b = \frac{M}{V} = \frac{1942}{908.94} = 2.14 \text{ g/cm}^3 \text{ thus } \gamma_b = \rho_b \times g = 20.99 \text{ kN/m}^3$$

Using the following relationship, dry density can then be determined:

$$\rho_b = \rho_d (1 + w)$$

Rearranging, we obtain:

$$\rho_d = \frac{\rho_b}{(1+w)} = \frac{2.14}{(1+0.137)} = 1.88 \text{ g/cm}^3, \text{ thus } \gamma_d = \rho_d \times g = 18.46 \text{ kN/m}^3$$

Hence, the bulk unit weight is 20.99 kN/m³ and the dry unit weight is 18.46 kN/m³.

Worked-out example 2.7: The results of a field compaction at various moisture contents using a heavy roller on a particular clayey fill were as shown in Table 2.3.

Table 2.3 Compaction data.

Water content [%]	Bulk density [g/cm³]
6.8	2.07
8.5	2.14
9.4	2.18
10.2	2.21
11.3	2.23
12.5	2.21
13.6	2.19

Plot the compaction curve (dry density/water content relationship), and on the same graph, plot the zero air voids (ZAV) and 5% air void lines. Assume a specific gravity of 2.7 for the clayey fill.

Solution: To plot the compaction curve, the dry density corresponding to the different water content and bulk density pairs can be determined with the relationship $\rho_b = \rho_d(1+w)$, as shown in Table 2.4.

Table 2.4 Compaction data, determination of dry density.

Water content [%]	Bulk density [g/cm³]	Dry density [g/cm³]
6.8	2.07	1.94
8.5	2.14	1.97
9.4	2.18	1.99
10.2	2.21	2.01
11.3	2.23	2.00
12.5	2.21	1.96
13.6	2.19	1.93

For determining zero air void line and 5% air voids, we can use the following relationships:

$$\rho_d = \frac{G_s \rho_w}{1+e} \Leftrightarrow \rho_d = \frac{G_s \rho_w}{1 + \dfrac{wG_s}{S_r}}$$

$$\therefore \rho_{d(ZAV,\, Sr=1)} = \frac{G_s \rho_w}{1 + wG_s}$$

$$A_v = \frac{V_a}{V} \text{ and } (1 - A_v) = 1 - \frac{V_a}{V} = \frac{V}{V} - \frac{V_a}{V} = \frac{V - V_a}{V} = \frac{V_w + V_s}{V} \text{ or } V = \frac{V_w + V_s}{(1 - A_v)}$$

$$\rho_d = \frac{\rho_b}{(1+w)} = \frac{M}{V(1+w)} = \frac{M_s + M_w}{V(1+w)}$$

By replacing V for the previous expression, the following expression is obtained:

$$\rho_d = \frac{(M_s + M_w)(1 - A_v)}{(V_w + V_s)(1+w)}$$

and we know that

$$\rho_s = \frac{M_s}{V_s} \text{ and } G_s = \frac{\rho_s}{\rho_w} \text{ then } V_s = \frac{M_s}{G_s \rho_w} \text{ and } \rho_w = \frac{W_w}{V_w} \text{ then } V_w = \frac{M_w}{\rho_w} \text{ or } V_w = \frac{wM_s}{\rho_w}$$

then,

$$\rho_d = \frac{(M_s + M_w)(1 - A_v)}{\left(\dfrac{wM_s}{\rho_w} + \dfrac{M_s}{G_s \rho_w}\right)(1+w)} = \frac{M_s \left(1 + \dfrac{M_w}{M_s}\right)(1 - A_v)}{\dfrac{M_s}{\rho_w}\left(w + \dfrac{1}{G_s}\right)(1+w)} = \frac{\rho_w(1+w)(1 - A_v)}{\left(w + \dfrac{1}{G_s}\right)(1+w)}$$

$$= \frac{\rho_w(1 - A_v)}{\dfrac{1}{G_s}(G_s w + 1)} = \frac{G_s \rho_w(1 - A_v)}{(G_s w + 1)}$$

$$\rho_d = \frac{G_s \rho_w(1 - A_v)}{(G_s w + 1)}$$

The values obtained using these relationships are shown in Table 2.5 and plotted in Figure 2.7.

Table 2.5 Data for plotting the ZAV and Av = 5% lines.

Water content [%]	Dry density (ZAV) [g/cm³]	Dry density (Av = 5%) [g/cm³]
6	2.32	2.21
7	2.27	2.16
8	2.22	2.11
9	2.17	2.06
10	2.13	2.02
11	2.08	1.98
12	2.04	1.94
13	2.00	1.90
14	1.96	1.86

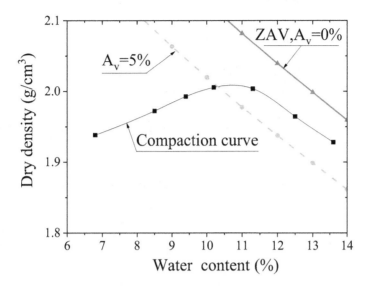

Figure 2.7 Compaction curve and curves representing the ZAV and A_v = 5% lines.

Worked-out example 2.8: Table 2.6 shows the results of standard compaction test on a sand/cement mixture having G = 2.70.

a.   Plot these results together with the zero air void line, and determine what percentage of air voids exists in the sample at OMC (i.e. A_{OMC}).

Table 2.6 Compaction data.

Water content [%]	5	8	10	12.5	16	20
Dry density [Mg/m³]	1.64	1.78	1.85	1.89	1.84	1.73

b.   The contract allows the field condition to vary as follows:

i. $A_{OMC} \pm 2.5\%$;
ii.   OMC + 3% and OMC – 4%

Determine the air void lines for the above conditions and plot them on the same graphs as in part a. Shade the zone in which the field product may fall.

Solution:
a.   For plotting the ZAV, we can use the following relationship:

$$\rho_d = \frac{G_s \rho_w}{1+e} \iff \rho_d = \frac{G_s \rho_w}{1 + \dfrac{wG_s}{S_r}}$$

$$\therefore \rho_{d(ZAV,\, Sr=1)} = \frac{G_s \rho_w}{1 + wG_s}$$

Table 2.7 shows the results obtained using the previous relationship, and these are shown in Figure 2.8.

Table 2.7 Data for plotting the ZAV.

Water content [%]	Dry density (ZAV) [g/cm³]
6	2.32
8	2.22
10	2.13
12	2.04
14	1.96
16	1.89
18	1.82
20	1.75
22	1.69

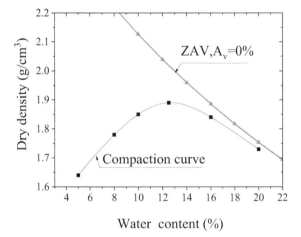

Figure 2.8 Compaction curve and curves representing the ZAV.

Using the relationship derived earlier $\rho_d = \dfrac{G_s \rho_w (1 - A_v)}{(G_s w + 1)}$ then, for OMC, we have

$\rho_d = 1.89 \text{ Mg/m}^3$

$w = 12.5\%$

$1.89 = \dfrac{2.7 \times 1 \times (1 - A_v)}{(2.7 \times 0.125 + 1)}$

$(1 - A_v) = 0.94 \text{ or } A_v = 0.064 \text{ or } A_v = 6.4\%$

b.   For the condition $A_{OMC} \pm 2.5\%$; then $A_v = 3.9\%$ and $A_v = 8.9\%$ (Fig. 2.9).

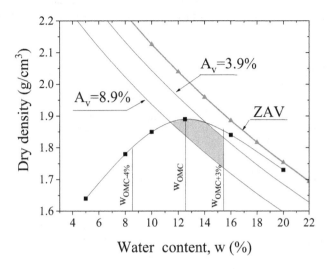

Figure 2.9 Compaction and allowed field specifications.

Worked-out example 2.9: The compaction characteristics of two industrial granular by-products and associated mixture (50% by weight of each component) were examined for evaluating their potential use in infrastructure projects. The compaction data is given in Table 2.8.

Plot the compaction data using the conventional ($\gamma_d - w$) and alternative ($e - e_w$) representation. The specific gravity of the granular waste materials A and B is given in Table 2.9.

Table 2.8 Compaction data of the waste materials considered.

Waste material A		Waste material A and B blend		Waste material B	
Water content [%]	Bulk density [g/cm³]	Water content [%]	Bulk density [g/cm³]	Water content [%]	Bulk density [g/cm³]
7	2.29	7	2.02	7	1.76
8.5	2.43	8.5	2.10	8.5	1.81
10.5	2.56	10.5	2.19	10.5	1.87
12	2.52	12	2.16	12	1.86
14	2.44	14	2.13	14	1.86

Table 2.9 Specific gravity of the waste materials considered.

Material tested	Specific gravity, G_s
Waste material A	3.2
Waste material A–B blend (50% by weight)	2.7
Waste material B	2.3

Solution: To plot the compaction data in the $(\gamma_d - w)$, the dry density corresponding to the different water content and bulk density pairs can be determined with the relationship:

$$\rho_d = \frac{\rho_b}{(1+w)}$$

and $\gamma_d = \rho_d g$ as shown in Table 2.10.

For plotting the different zero air void (ZAV) lines, the following relationship can be used (Table 2.11):

$$\rho_d = \frac{G_s \rho_w}{1+e} \Leftrightarrow \rho_d = \frac{G_s \rho_w}{1 + \dfrac{wG_s}{S_r}}$$

$$\therefore \rho_{d(ZAV,\ Sr=1)} = \frac{G_s \rho_w}{1 + wG_s}$$

Then the results can be plotted in terms of $(\gamma_d - w)$ as shown in Figure 2.10.

Table 2.10 Compaction data, determination of dry unit weight.

Waste material A

Water content [%]	Bulk density [g/cm³]	Dry density [g/cm³]	Dry unit weight [kN/m³]
7	2.29	2.14	21
8.5	2.43	2.24	22
10.5	2.56	2.32	22.7
12	2.52	2.25	22.1
14	2.44	2.14	21

Waste material A and B blend

Water content [%]	Bulk density [g/cm³]	Dry density [g/cm³]	Dry unit weight [kN/m³]
7	2.02	1.89	18.5
8.5	2.10	1.94	19
10.5	2.19	1.98	19.4
12	2.16	1.93	18.9
14	2.13	1.87	18.3

Waste material B

Water content [%]	Bulk density [g/cm³]	Dry density [g/cm³]	Dry unit weight [kN/m³]
7	1.76	1.64	16.1
8.5	1.81	1.67	16.4
10.5	1.87	1.69	16.6
12	1.86	1.66	16.3
14	1.86	1.63	16

Table 2.11 Data for plotting the ZAV.

Water content [%]	Dry density (ZAV, Gs = 3.2) [g/cm³]	Dry density (ZAV, Gs = 2.7) [g/cm³]	Dry density (ZAV, Gs = 2.3) [g/cm³]
6	2.68	2.32	2.02
7	2.61	2.27	1.98
8	2.55	2.22	1.94
9	2.48	2.17	1.91
10	2.42	2.13	1.87
11	2.37	2.08	1.84
12	2.31	2.04	1.80
13	2.26	2.00	1.77
14	2.21	1.96	1.74

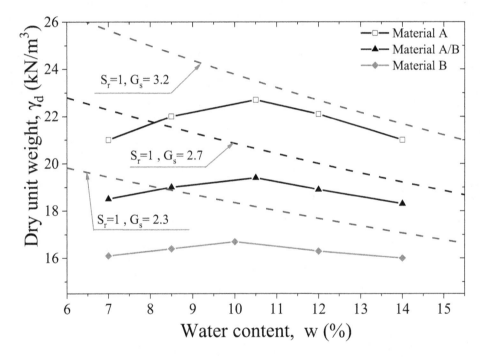

Figure 2.10 Compaction data plotted in terms of (γ_d – w).

To plot the compaction data in the (e – e_w), the void ratio and water ratio corresponding to the different water content and dry unit weights pairs can be determined with the relationship (Table 2.12), and the results are illustrated in Figure 2.11.

$$e = \frac{\gamma_w G_s}{\gamma_d} - 1$$

$$e_w = eS_r = wG_s$$

Table 2.12 Compaction data, determination of e and e_w.

Waste material A

Water content [%]	e_w	Dry unit weight [kN/m³]	e
7	0.22	21	0.49
8.5	0.27	22	0.43
10.5	0.34	22.7	0.38
12	0.38	22.1	0.42
14	0.45	21	0.49

Waste material A and B blend

Water content [%]	e_w	Dry unit weight [kN/m³]	e
7	0.19	18.5	0.43
8.5	0.23	19	0.39
10.5	0.28	19.4	0.36
12	0.32	18.9	0.40
14	0.38	18.3	0.45

Waste material B

Water content [%]	e_w	Dry unit weight [kN/m³]	e
7	0.16	16.1	0.40
8.5	0.20	16.4	0.38
10.5	0.24	16.6	0.36
12	0.28	16.3	0.38
14	0.32	16	0.41

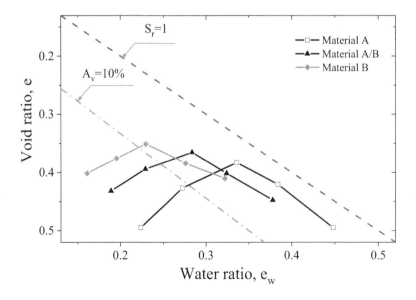

Figure 2.11 Compaction data plotted in terms of (γ_d–w).

References

Knappett, J. & Craig, R.F. (2012) *Craig's Soil Mechanics*, 8th edition. CRC Press, ISBN:9780415561266, p. 584.

Standards Australia (2007) *Guidelines on Earthworks for Commercial and Residential Developments*. Australian Standard AS3798-2007, Standards Australia, Sydney.

Standards Australia (2017) *Methods of Testing Soils for Engineering Purposes– Soil Compaction and Density Tests – Determination of the Dry Density/Moisture Content Relation of a Soil Using Standard Compactive Effort*. Australian Standard AS1289.5.1.1-2017, Standards Australia, Sydney.

Chapter 3

Effective stress concepts

This chapter introduces the state of stress in a soil element and the effective stress concept.

3.1 State of stress in the soil element

The state of stress in the soil element can be defined based on total stress (σ), pore pressure (u) and effective stress (σ').

Total stress (σ) is the stress acting on a soil element located at a certain depth and is defined as a function of the bulk unit weight of the materials resting above and their associated depth. For instance, for a homogeneous ground having a unit weight of 20 kN/m^3, a soil element located 10 m deep would have a total stress of 200 kPa.

Pore pressure (u) is the pressure developed in the pores of soil that are filled with fluid, either water or air, or a combination of both. These fluids do not offer resistance to static shear forces, but they can support vertical pressures.

Effective stress (σ') the effective stress is dependent on the total stress (σ) and on the pore air (u_a) and pore water (u_w) pressures. For instance, in stability and compressibility problems, changes in effective stress govern the changes in volume and shear strength.

3.2 Effective stress in saturated soils (or Terzaghi effective stress)

Terzaghi (1943) demonstrated through a series of laboratory experiments that for saturated soils (e.g. degree of saturation of 1), effective stress can be defined using an empirical relationship, as follows:

$$\sigma' = \sigma - u_w \tag{3.1}$$

Where σ' is the effective stress, σ is the total stress and u_w is the pore water pressure. This is valid for evaluating soil compressibility and shear strength provided that the area of contact between particles is small and that the compressibility of the individual particles is small compared with that of the soil skeleton.

3.3 Effective stress in partially saturated soils

For the cases where voids are filled with air and water, surface tension occurs at the air/ water interfaces. As a result, the pressures in the water and air phase are not the same.

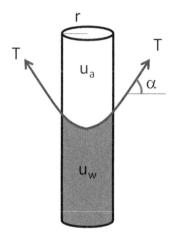

Figure 3.1 Capillary tube illustration, with *r* being the radius of the tube.

For instance, consider the capillary tube displayed in Figure 3.1. For equilibrium the following holds:

$$u_a \, \pi r^2 = u_w \, \pi r^2 + 2 \, \pi r \sin \alpha \, T \tag{3.2}$$

Where T is the surface tension expressed per unit length of air/water interface. Rearranging we can obtain:

$$(u_a - u_w) = \frac{2T \sin \alpha}{r} \tag{3.3}$$

The meniscus radius depends on the pore size. For instance, for sands the pressure difference between air and water phases is relatively small (e.g. dozens of kPa), but for clay soils having smaller pore sizes $(u_a - u_w)$ can reach thousands of kPa. These concepts as well as the effective stress for unsaturated soils will be covered in more detail in Chapter 14.

In this chapter, the focus is on the Terzaghi effective stress, henceforth referred to simply as effective stress.

3.4 State of stress and groundwater

The pore water pressure is one of the key parameters that governs the behavior of soil. The evaluation of pore water pressure is critical for two main conditions, i.e. static (no flow) and seepage (water flow). In this chapter, the focus is on the static conditions, whereas seepage is covered in Chapter 4.

The pressure of pore water is measured relative to the atmospheric pressure, and the level at which the pressure is atmospheric (taken as 0 for simplicity) is defined as the groundwater table (GWT) or phreatic surface. Below the water table, the soil is typically assumed to be saturated (degree of saturation = 1). Immediately above the GWT, the soil remains saturated with water due to capillary stresses, which hold the water below atmospheric pressure, referred to as the capillary fringe. Above this level, there is a partially saturated zone or vadose zone, where water is held by surface tension and adsorption (Fig. 3.2). While level of GWT typically varies according to climatic conditions (e.g. rainfall), it can also vary because of construction operations (e.g. excavations).

There are cases in which a perched water table can occur in discrete locations, and it's typically associated with an aquitard or aquiclude (water reservoirs having very low permeability or impermeable), as illustrated in Figure 3.3. Artesian conditions may also occur if a relatively high permeability layer is confined by an aquitard or aquiclude. In these conditions, the water pressure in the artesian layer is governed by a higher water table, i.e. potentiometric surface (Fig. 3.4).

Consider a soil element located at a certain depth (z) and a water table at surface level. The total vertical stress (σ_v) at depth z is equal to the weight of all the material (solids and

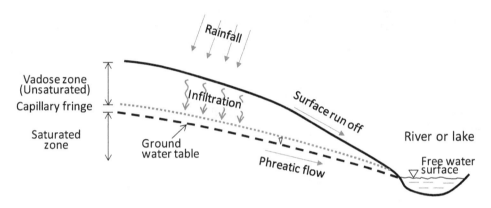

Figure 3.2 Illustration of a typical profile illustrating the occurrence of groundwater (modified after Whitlow, 1995).

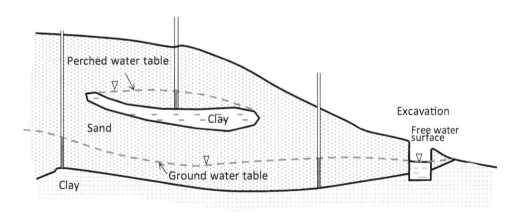

Figure 3.3 Illustration of a typical profile illustrating the occurrence of perched water table (modified after Whitlow, 1995).

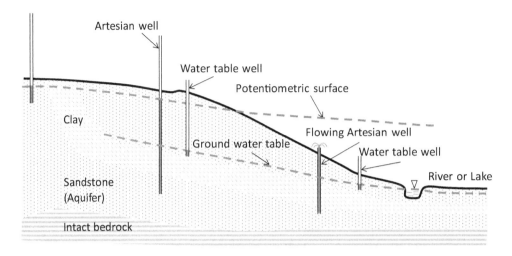

Figure 3.4 Illustration of a typical profile illustrating the occurrence of artesian conditions.

water) per unit areas above that depth, whereas the pore water pressure (u) at depth z will be hydrostatic, or

$$\sigma_v = \gamma_{sat} z \tag{3.4}$$

$$u = \gamma_w z \tag{3.5}$$

Where γ_{sat} is the saturated unit weight and γ_w is the unit weight of the water. Thus, the effective stress at a certain depth z can be computed as follows:

$$\sigma_v' = \sigma_v - u \quad \text{or}$$
$$\sigma_v' = (\gamma_{sat} - \gamma_w) z \tag{3.6}$$

Worked-out example 3.1: A layer of saturated cohesive soil has a water content of 54%, and the specific gravity of solid particles of the soil is 2.78. Calculate the vertical effective stress at a depth of 15.4 m, if

a.   the water table is at the ground surface, or
b.   the water table is at a depth of 6 m below the ground surface and the soil above the water table is saturated.

Solution:
a.   For determining the effective stress at 15.4 m, we first need to calculate the saturated unit weight. As the water table is at ground surface, the soil is fully saturated and

$$S_r e = w G_s \quad \text{or} \quad e = w G_s$$

Then the bulk (saturated) unit weight can be determined as follows:

$$\gamma_b = \frac{W}{V} = \frac{G_S + S_r e}{1+e}\gamma_w$$

$$= \left(\frac{2.78 + 1(0.54 \times 2.78)}{1 + (0.54 \times 2.78)}\right)9.81$$

$$\gamma_b = 16.79\,\text{kN/m}^3$$

Total vertical stress: $\sigma_v = \gamma_{sat} z = 16.79 \times 15.4 = 258.57\,\text{kPa}$
Pore water pressure: $u = \gamma_w z = 9.81 \times 15.4 = 151.1\,\text{kPa}$
Effective vertical stress: $\sigma_v' = \sigma_v - u = 258.57 - 151.1 = 107.5\,\text{kPa}$

b.   For a water table located at a depth of 6 m depth, as the soil remains fully saturated, the total stress is the same, but the pore water pressure at a depth of 15.4 m is smaller, as there is only 9.4 m of hydrostatic pressure at that depth, or

Total vertical stress: $\sigma_v = \gamma_{sat} z = 16.79 \times 15.4 = 258.57\,\text{kPa}$
Pore water pressure: $u = \gamma_w z = 9.81 \times (15.4 - 6) = 92.2\,\text{kPa}$
Effective vertical stress: $\sigma_v' = \sigma_v - u = 258.57 - 92.2 = 166.4\,\text{kPa}$

Worked-out example 3.2: For the soil profile shown in Figure 3.5,

a.   Draw diagrams indicating the total and effective stress, and pore water pressure on the horizontal planes to a depth of 11 m below the ground surface. (Assume sand above GWT is dry.)
b.   How do the pore water pressure and effective stress change if the sand above the GWT remains saturated with capillary moisture?

   The saturated unit weights of sand and clay are 20 kN/m³ and 18.5 kN/m³ respectively, and the dry unit weight of sand is 17 kN/m³.

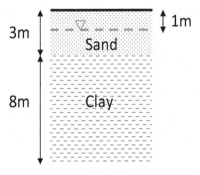

Figure 3.5 Soil profile.

Solution:

a. For drawing the total and effective stress, and pore water pressure profile with depth, points at different depths (0, 1, 3 and 11 m) are considered. Results are tabulated in Table 3.1.

As the stresses increase linearly with depth, the diagrams can be drawn as shown in Figure 3.6.

b. For the case in which sand is saturated with capillary moisture, the first meter of the pore water pressure profile changes and the total vertical and effective stress profiles shift by

Table 3.1 Total and effective stress and pore water pressure results.

Depth (m)	Total vertical stress	Pore water pressure	Effective vertical stress
0	0	0	0
1	$\sigma_v = \gamma_{dry} z = 1 \times 17 = 17\,kPa$	$u = \gamma_w z = 0\,kPa$	$\sigma_v' = \sigma_v - u$
			$\sigma_v' = 17 - 0 = 17\,kPa$
3	$\sigma_v = \gamma_{dry} z + \gamma_{sat} z$	$u = \gamma_w z$	$\sigma_v' = \sigma_v - u$
	$\sigma_v = 17 \times 1 + 20 \times 2 = 57\,kPa$	$u = 9.81 \times 2 = 19.6\,kPa$	$\sigma_v' = 57 - 19.6$
			$\sigma_v' = 37.4\,kPa$
11	$\sigma_v = \gamma_{dry} z + \left(\gamma_{sat} z\right)_{sand} + \left(\gamma_{sat} z\right)_{clay}$	$u = \gamma_w z$	$\sigma_v' = \sigma_v - u$
	$\sigma_v = 17 \times 1 + 20 \times 2 + 18.5 \times 8$	$u = 9.81 \times 10 = 98.1\,kPa$	$\sigma_v' = 205 - 98.1$
	$\sigma_v = 205\,kPa$		$\sigma_v' = 106.9\,kPa$

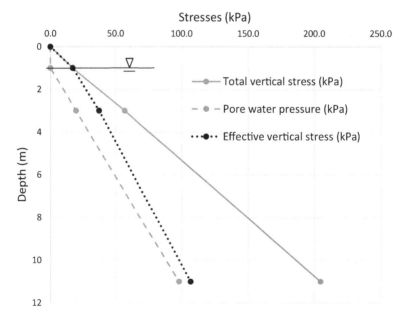

Figure 3.6 Total and effective stress and pore water pressure diagrams.

3 kPa. This corresponds to an increase in unit weight of the sand from 17 kN/m³ (dry condition) to 20 kN/m³ (saturated condition) over 1 m above the GWT where capillary moisture holds. The results can be computed similarly as shown in Table 3.2 and illustrated in Figure 3.7.

Where z_0 represents the depth from the GWT to the point of interest.

Table 3.2 Total and effective stress and pore water pressure results.

Depth (m)	Total vertical stress (kPa)	Pore water pressure (kPa)	Effective vertical stress (kPa)
0	0	$u = \gamma_w(-z_0)$ $= -9.81 \times 1 = -9.8\,kPa$	9.8
1	20	0	20
3	60	19.6	40.4
11	208	98.1	109.9

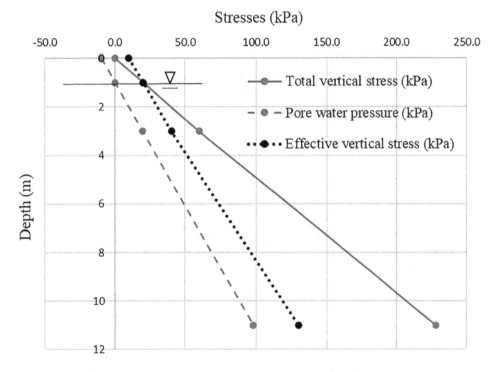

Figure 3.7 Total and effective stress and pore water pressure diagrams.

Worked-out example 3.3: The surface of a deposit of clay is horizontal and the water table is located at the surface.

a. Compute the pore water pressure at point A located 1.8 m below the ground surface;
b. Dewatering operations have commenced in the vicinity of this site, and groundwater table has dropped to 4 m below the ground surface. Compute the pore water pressure of point A, assuming that the clay above the water table remains saturated with capillary moisture;
c. Does the drop in water table change the effective and total stresses at point A, and if so, by how much?

Solution:
a. The pore water pressure at 1.8 m below the ground surface can be computed as follows:

$$u = \gamma_w z$$
$$u = 9.81 \times 1.8 = 17.7 \, \text{kPa}$$

b. If the water table dropped to a deeper location and the soil remains saturated with capillary moisture, then the pore water pressure at point A can be computed as follows:

$$u = \gamma_w (-z_0)$$
$$u = 9.81 \times (-(4-1.8)) = -21.58 \, \text{kPa}$$

c. While the total stress remains unchanged because it depends on the saturated unit weight of the soil ($\sigma_v = \gamma_{sat} z$), the effective stress will vary because there is a variation in the pore water pressure, or

$$\sigma_v' = \sigma_v - u$$

Thus, if the GWT level is lowered to a higher depth, the pore water pressure will be smaller, and the effective stress will increase. The amount of this increase is directly associated with the magnitude of change of pore water pressure. In this exercise, the variation of pore water pressure can be computed considering the initial and final pore water pressure values, as follows:

$$\Delta \sigma_v' = \Delta u$$
$$\Delta u = 17.74 - (-21.58)$$
$$\Delta u = 39.32 \, \text{kPa}$$

Worked-out example 3.4: A large excavation was opened in a stratum of stiff clay with a saturated unit weight of 17 kN/m³. When the depth of the excavation reached 4.5 m, the bottom cracked and was flooded from below by a mixture of sand and water. Subsequent borings showed that the clay was underlain by a bed of sand with its surface at a depth of 12.5 m.

Compute the water head in the sand layer and the water level in a well located in sand layer if this had been installed before the excavation started.

Solution: As the bottom of the excavation cracked, that indicates that the pore water pressure (water head) in the sand layer exceeded the vertical stress at the separation between the clay and sand layer (Point A, Fig. 3.8), or effective stress = 0.

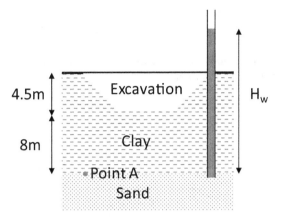

Figure 3.8 Ground profile diagram.

Thus, considering the equilibrium conditions at point A, just before failure, then:

$$\sigma_v' = \sigma_v - u \; ; \; \sigma_v' = 0$$

$$\sigma_v = u$$

$$(12.5 - 4.5)\gamma_{sat} = \gamma_w H_w \text{ with water head} = \gamma_w H_w$$

$$\gamma_w H_w = 8 \times 17$$

$$\gamma_w H_w = 136 \, \text{kPa}$$

Using the same equilibrium relationship, the water level in the well located in the sand layer can be computed as follows:

$$136 = \gamma_w H_w$$

$$H_w = \frac{136}{9.81}$$

$$H_w = 13.7 \, \text{m}$$

The water head at point A is 136 kPa, and the water level in the well located in the sand layer is 13.7 m.

Worked-out example 3.5: A site investigation conducted for a large excavation detected a 10-m thick layer of uniform clay resting on a 3-m thick layer of sand, which was underlain by bedrock. The water level was at ground level, and the sand had a head of 2 m above the top of the clay. The saturated unit weights of the clay and sand were 19.8 kN/m³ and 21 kN/m³, respectively.

a.   Draw the profile of total and effective stress and pore water pressure to the depth of the bedrock (13 m deep).
b.   Calculate the depth to which the excavation can be taken before ground heave occurs.

Solution:

a.   To draw the profile of stresses and pore water pressure, the magnitude of the stresses must be determined at different depths (Fig. 3.9), as shown in Table 3.3.

Once the values are computed, the profile of stresses and pore water pressure with depth can be represented as illustrated in Figure 3.10.

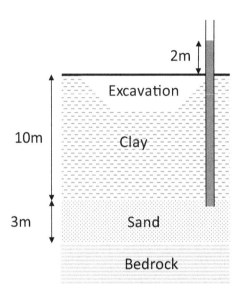

Figure 3.9 Ground profile diagram.

Table 3.3 Total and effective stress and pore water pressure results.

Depth (m)	Total vertical stress (kPa)	Pore water pressure (kPa)	Effective vertical stress (kPa)
0	$\sigma_v = \gamma_{sat} z = 19.8 \times 0 = 0$	$u = \gamma_w z = 0$	$\sigma_v' = \sigma_v - u = 0$
10 (clay)	$\sigma_v = \gamma_{sat} z = 19.8 \times 10 = 198$	$u = \gamma_w z = 10 \times 9.81$ $= 98.1$	$\sigma_v' = \sigma_v - u$ $\sigma_v' = 198 - 98.1$ $\sigma_v' = 99.9$
10 (sand)	$\sigma_v = \gamma_{sat} z = 19.8 \times 10 = 198$	$u = \gamma_w z = 12 \times 9.81$ $= 117.72$	$\sigma_v' = \sigma_v - u$ $\sigma_v' = 198 - 117.72$ $\sigma_v' = 80.3$
13	$\sigma_v = (\gamma_{sat} z)_{clay} + (\gamma_{sat} z)_{sand}$ $\sigma_v = 198 + 21 \times 3$ $\sigma_v = 261$	$u = \gamma_w z = (12 + 3) \times 9.81$ $= 147.15$	$\sigma_v' = \sigma_v - u$ $\sigma_v' = 261 - 147$ $\sigma_v' = 113.85$

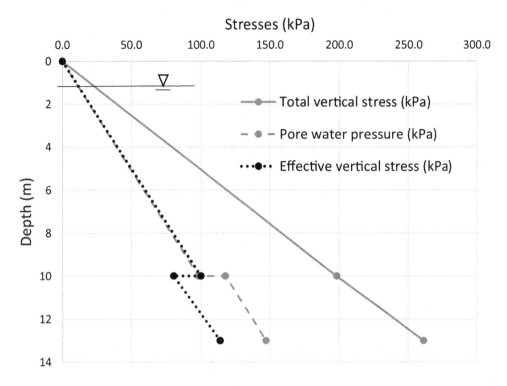

Figure 3.10 Total and effective stress and pore water pressure diagrams.

b. Ground heave at a certain excavation level will occur when the effective stress in the clay layer drops to 0 or

$$\sigma_v' = \sigma_v - u \; ; \; \sigma_v' = 0$$

$$\sigma_v = u$$

$$(10 - x)\gamma_{sat} = \gamma_w H_w \text{ with water head sand layer} = \gamma_w H_w$$

$$\gamma_w H_w = 12 \times 9.81$$

$$\gamma_w H_w = 117.7 \, \text{kPa}$$

Using the same equilibrium relationship, the depth of excavation before heave occurs can be computed as follows:

$$(10 - x)19.8 = 117.7$$

$$x = 10 - \frac{117.7}{19.8}$$

$$x = 4.1 \, \text{m}$$

The ground heave will occur for an excavation level greater than 4.1 m deep.

Worked-out example 3.6: A construction of a new building's basement is proposed for a site having a ground profile as shown in Figure 3.11. To enable the commencement of construction work, a dewatering system is installed on site, which is able to lower the groundwater level to 2 m below ground level. Calculate the total and effective stress as well as pore water pressure profile for the ground before and after installation of the dewatering system. Draw the associated profile with depth for the two conditions. The saturated unit weights of sand and clay are 20 kN/m³ and 18.5 kN/m³, respectively, and the dry unit weight of sand is 17 kN/m³.

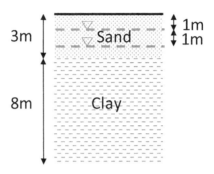

Figure 3.11 Ground profile diagram.

Solution: The total, effective and pore water pressure profiles of the ground for a GWL of 1 m depth have been computed earlier in worked-out example 3.2. For the GWL of 2 m depth, results are shown in Table 3.4.

Table 3.4 Total and effective stress and pore water pressure results after dewatering system installation.

Depth (m)	Total vertical stress (kPa)	Pore water pressure (kPa)	Effective vertical stress (kPa)
0	0	0	0
1	$\sigma_v = \gamma_{dry} z = 1 \times 17 = 17\,kPa$	$u = \gamma_w z = 0\,kPa$	$\sigma_v' = \sigma_v - u$
			$\sigma_v' = 17 - 0 = 17\,kPa$
2	$\sigma_v = \gamma_{dry} z$	$u = \gamma_w z$	$\sigma_v' = \sigma_v - u$
	$\sigma_v = 17 \times 2 = 34\,kPa$	$u = 0\,kPa$	$\sigma_v' = 34 - 0 = 34\,kPa$
3	$\sigma_v = \gamma_{dry} z + \gamma_{sat} z$	$u = \gamma_w z$	$\sigma_v' = \sigma_v - u$
	$\sigma_v = 17 \times 2 + 20 \times 1 = 54\,kPa$	$u = 9.81 \times 1 = 9.8\,kPa$	$\sigma_v' = 54 - 9.8$
			$\sigma_v' = 44.2\,kPa$
11	$\sigma_v = \gamma_{dry} z + \left(\gamma_{sat} z\right)_{sand} + \left(\gamma_{sat} z\right)_{clay}$	$u = \gamma_w z$	$\sigma_v' = \sigma_v - u$
	$\sigma_v = 54 + 18.5 \times 8$	$u = 9.81 \times 9 = 88.3\,kPa$	$\sigma_v' = 202 - 88.3$
	$\sigma_v = 202\,kPa$		$\sigma_v' = 113.7\,kPa$

The total, effective stress and pore water pressure profiles for the two cases are shown in Figure 3.12. Note that it was assumed that the sand above GWL is dry in both conditions.

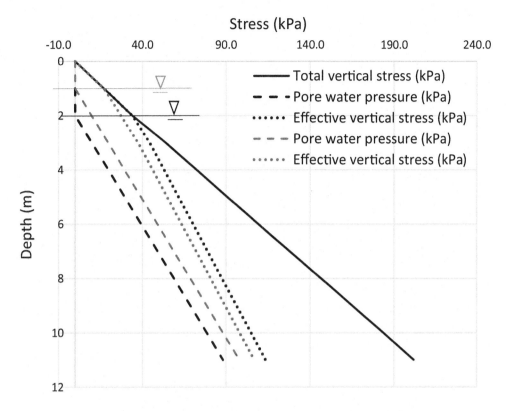

Figure 3.12 Total and effective stress and pore water pressure diagrams (grey series represent conditions after the installation of the dewatering system).

References

Terzaghi, K. (1943) *Theoretical Soil Mechanics*. John Wiley & Sons, London, p. 503.
Whitlow, R. (1995) *Basic Soil Mechanics*, 3rd edition. Longman Scientific & Technical, London, p. 559.

Chapter 4

Permeability and seepage

This chapter introduces different aspects related to the flow of water through soils. Soils are permeable materials that allow water to flow through their interconnected pores. The study of the flow through soil is vital for geotechnical applications such as excavations located at a depth greater than the groundwater level where dewatering is required or determines the yield of wells. In addition, the flow establishes a pattern of pore water pressures that may lead to the development of critical stability conditions or induce volume changes in soil that could result in settlement or ground heave.

4.1 Permeability

Permeability is a measure of the rate at which the fluid passes through a porous medium. For water, the apparent velocity of the flow (v) can be related to the hydraulic gradient (i) and permeability (k) via Darcy's law, as follows:

$$v = k\frac{dh}{dl} = ki$$

(4.1)

where dh represents the hydraulic head variation along the flow path dl.

Eq. 4.1 can be rewritten to account for the rate of flow (q) across a cross-sectional area A.

$$q = Aki$$

(4.2)

Thus, the rate of water flow depends upon the permeability that is in turn influenced by the density and viscosity of the fluid, turbulence of flow, porosity of the soil, shape and arrangement of soil particles, degree of saturation and thickness of adsorbed layers (fine-grained soils).

4.1.1 Permeability estimation

Kozeny (1927) and Carman (1956) proposed and demonstrated that for laminar flow in saturated soils, permeability may be estimated as a function of different coefficients to consider the shape of the pores (k_0), tortuosity (K_T), particle surface area (S_s), porosity (n), unit weight of the water (γ_w) and viscosity of the water (η).

$$k = \frac{1}{k_0 k_T S_s^2} \frac{n^3}{(1-n)^2} \frac{\gamma_w}{\eta}$$

(4.3)

Typical values of permeability for soils are given in Table 4.1. However, these are only indicative and may vary considerably if even small quantities of fine materials are present.

4.1.2 Measurement of permeability in laboratory

There are typically two types of tests that can be adopted for the determination of the coefficient of permeability of soil. For coarse-grained materials, a constant head permeameter (Fig. 4.1a) is used, whereas for fine-grained materials, a falling head permeameter is selected (Fig. 4.1b).

Table 4.1 Typical values of permeability.

Permeability (m/s)										
1	10^{-1}	10^{-2}	10^{-3}	10^{-4}	10^{-5}	10^{-6}	10^{-7}	10^{-8}	10^{-9}	10^{-10}
Clean gravels										
		Clean sands								
					Silt					
				Fissured clay						
							Intact clay			

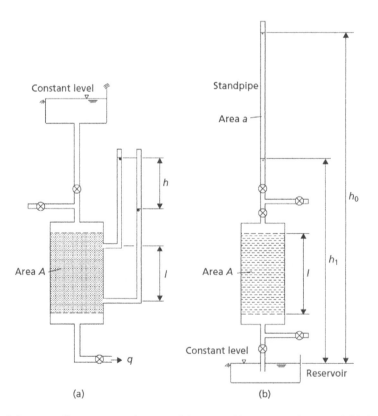

(a)   (b)

Figure 4.1 Schematic illustrations of permeability test (a) constant head and (b) falling head (after Knappett and Craig, 2012 reproduced with permission from CRC Press).

4.1.3 Measurement of permeability in the field using a pumping test

Case 1: unconfined aquifer

The hydraulic gradient can be expressed as $i = dh/dr$, whereas the area through which flow takes place is $A = 2\pi rH$ (OW1 and OW2 in Fig. 4.2). Applying Darcy's Law,

$$(q = k\,i\,A)$$

Then $q = k\dfrac{dh}{dr}2\pi\,R\,H$

Upon integration using appropriate limits (R_1 and R_2 and H_1 and H_2):

$$q\int_{R1}^{R2}\frac{dR}{R} = \pi k\int_{H1}^{H2}2hdh \ \text{ or}$$

$$q\,ln\frac{R_2}{R_1} = \pi Hk\left(H_2^2 - H_1^2\right)$$

Rearranging for permeability, then

$$k = \frac{q\,ln\left(\dfrac{R_2}{R_1}\right)}{\pi\left(H_2^2 - H_1^2\right)} \tag{4.4}$$

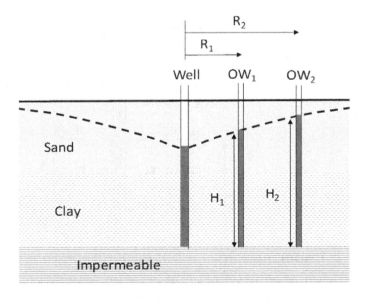

Figure 4.2 Well pumping test in a unconfined aquifer.

Case 2: confined aquifer

For a confined stratum of thickness H (Fig. 4.3), the area through which water flows through is $A = 2\pi r H$, where H is a constant (thickness of the confined layer), applying Darcy's law ($q = k i A$), the flow can be obtained as follows:

$$q = k\frac{dh}{dr}2\pi r H$$

On integrating using appropriate limits (e.g. R_1 and R_2 and H_1 and H_2),

$$q\int_{R1}^{R2}\frac{dR}{R} = 2\pi Hk\int_{H1}^{H2}dh$$

$$q\,ln\frac{R_2}{R_1} = 2\pi Hk\,(H_2 - H_1)$$

Rearranging, then

$$k = \frac{q\,ln\left(\dfrac{R_2}{R_1}\right)}{2\pi H\,(H_2 - H_1)} \tag{4.5}$$

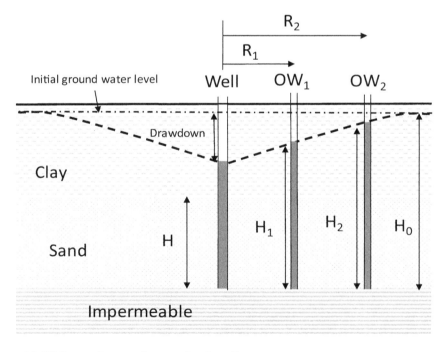

Figure 4.3 Well pumping test in a confined aquifer.

4.1.4 Permeability of stratified soils

When the ground profile consists of a number of different layers of soil having different permeability values, the overall permeability is not the same in the horizontal and vertical directions.

Horizontal flow

For horizontal flow (Fig. 4.4), the equivalent permeability can be computed as follows:

$$k = \frac{q}{Ai}$$

$$k = \frac{q_1 + q_2 + q_3}{(A_1 + A_2 + A_3) \times (i_1 + i_2 + i_3)}$$

As the $q = Aki$, then

$$k = \frac{A_1 k_1 i_1 + A_2 k_2 i_2 + A_3 k_3 i_3}{(A_1 + A_2 + A_3) \times (i_1 + i_2 + i_3)}$$

As the hydraulic gradient in each layer is the same, the previous equation can be simplified into:

$$k_H = \frac{D_1 k_1 + D_2 k_2 + D_3 k_3}{D_1 + D_2 + D_3} \tag{4.6}$$

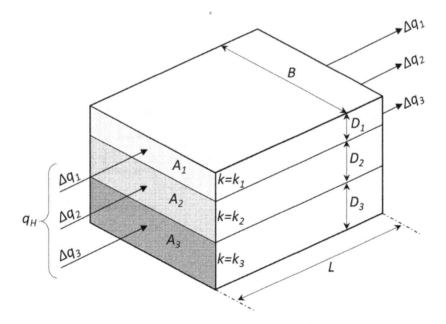

Figure 4.4 Horizontal flow through stratified soils (modified after Whitlow, 1995).

Vertical flow

For vertical flow (Fig. 4.5), the equivalent permeability can be computed as follows:

$$k = \frac{q}{Ai}$$

The total flow is equal to the flow in each layer, whereas the total head loss is equal to the sum of the losses in the different soil layers.

$$k = \frac{q_y}{A} \times \frac{(l_1 + l_2 + l_3)}{(h_1 + h_2 + h_3)}$$

$$k = \frac{q_y}{A} \times \frac{(l_1 + l_2 + l_3)}{(i_1 l_1 + i_2 l_2 + i_3 l_3)}$$

$$k = \frac{q_y}{A} \times \frac{(l_1 + l_2 + l_3)}{\left(\dfrac{q_1 l_1}{A_1 k_1} + \dfrac{q_2 l_2}{A_2 k_2} + \dfrac{q_3 l_3}{A_3 k_3} \right)}$$

Upon simplification, the following relationship can be obtained.

$$k_V = \frac{D_1 + D_2 + D_3}{\dfrac{D_1}{k_1} + \dfrac{D_2}{k_2} + \dfrac{D_3}{k_3}} \tag{4.7}$$

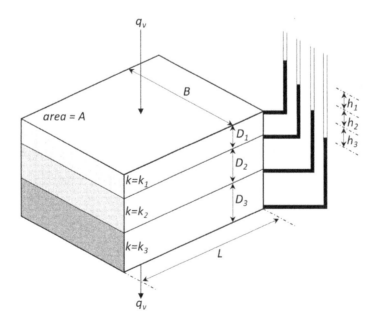

Figure 4.5 Vertical flow through stratified soils (modified after Whitlow, 1995).

4.2 Seepage theory

4.2.1 Two-dimensional flow

Consider a soil element of dimension dx, dy and dz in an incompressible fluid having velocities v_x and v_z. The quantity flowing into element equals the quantity flowing out of element (Fig. 4.6), as follows:

$$V_x \cdot dz + V_z \cdot dx = \left[V_x + \frac{\partial V_x}{\partial x} \cdot dx\right] dz + \left[V_z + \frac{\partial V_z}{\partial z} \cdot dz\right] dx \qquad (4.8)$$

For steady state condition the net inflow should be zero, thus,

$$\frac{\partial v_x}{\partial x} dx dz + \frac{\partial v_z}{\partial z} dz dx = 0$$

$$\frac{\partial v_x}{\partial x} + \frac{\partial v_z}{\partial z} = 0$$

4.2.2 Application to groundwater flow using Darcy's law

The two-dimensional flow solutions may be applied to problems concerning the flow of water through the pores of the soil. Considering Darcy's law, the apparent velocities in the horizontal and vertical direction, v_x and v_z, respectively, can be computed as follows:

$$v_x = k \cdot \frac{\partial h}{\partial x} \qquad (4.9)$$

$$v_z = k \cdot \frac{\partial h}{\partial z}$$

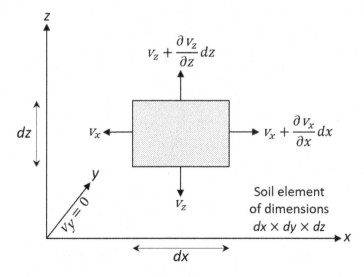

Figure 4.6 General two-dimensional flow condition (modified after Whitlow, 1995).

Then for steady-state conditions in an incompressible fluid, we have:

$$0 = \frac{\partial^2 h}{\partial x^2} + \frac{\partial^2 h}{\partial z^2} \qquad (4.10)$$

4.2.3 Properties of flow nets

Direction of flow is represented by *Flow Lines* (stream lines). Variation of pressure (head) across the flow is represented by *Equipotential Lines*. Mapping of flow lines and equipotential lines together is called a *Flow Net*. The Flow Lines (stream lines) and equipotential Lines are orthogonal (Fig. 4.7), and boundaries of no flow are typically taken as flow lines, whereas submerged permeable boundaries are equipotential lines.

4.2.4 Anisotropic flow

If $k_x \neq k_z$, then Eq. 4.10 can be expressed in the Laplace form as follows:

$$k_x \frac{\partial^2 h}{\partial x^2} + k_z \frac{\partial^2 h}{\partial z^2} = 0$$

Considering $x_1 = x\sqrt{\dfrac{k_y}{k_x}}$ and since $\dfrac{\partial h}{\partial x} = \dfrac{\partial h}{\partial x_1} \cdot \dfrac{\partial x_1}{\partial x} = \dfrac{\partial h}{\partial x_1}\sqrt{\dfrac{k_y}{k_x}}$ and $\dfrac{\partial^2 h}{\partial x^2} = \sqrt{\dfrac{k_y}{k_x}} \cdot \dfrac{\partial^2 h}{\partial x_1^2} \cdot$

$\sqrt{\dfrac{k_y}{k_x}} = \dfrac{\partial^2 h}{\partial x_1^2} \cdot \dfrac{k_y}{k_x}$

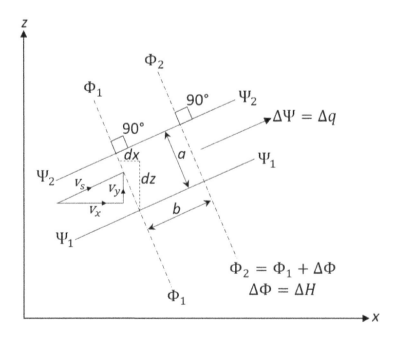

Figure 4.7 Graphical representation of flow (modified after Whitlow, 1995).

From the Laplace equation:

$$k_x \left[\frac{\partial^2 h}{\partial x_1^2} \cdot \frac{k_y}{k_x} \right] + k_y \cdot \frac{\partial^2 h}{\partial y^2} = 0 \text{ or}$$

$$\frac{\partial^2 h}{\partial x_1^2} + \frac{\partial^2 h}{\partial y^2} = 0$$

Therefore, by making the x-axis transformed by the factor $\sqrt{\frac{k_y}{k_x}}$, we can now do the flow net analysis as before, where $k_f = k_x \sqrt{\frac{k_y}{k_x}} = \sqrt{k_x \cdot k_y}$.

4.2.5 Flowline deflection at anisotropic boundary

Darcy's Law in Zone 1 (Fig. 4.8):

$$q_1 = A_1 \cdot k_1 \cdot i_1 = (\cos \beta_1) \cdot k_1 (h_1 - h_2) / (\sin \beta_1)$$

Darcy's Law in Zone 2 (Fig. 4.8):

$$q_2 = A_2 \cdot k_2 \cdot i_2 = (\cos \beta_2) \cdot k_2 (h_1 - h_2) / (\sin \beta_2)$$

But along the same flow channel,

$q_1 = q_2$ (continuity of flow)

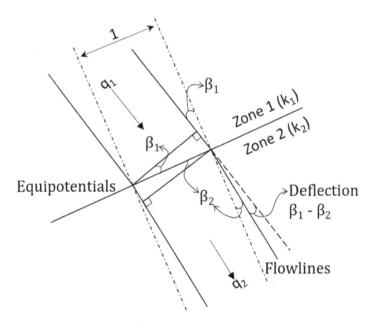

Figure 4.8 Graphical representation of flow in anisotropic media (modified after Whitlow, 1995).

Finally, $\dfrac{\tan \beta_1}{\tan \beta_2} = \dfrac{k_1}{k_2}$

4.2.6 Determination of seepage flow using flow nets

Considering a flow net to describe the seepage flow under a sheet pile (Fig. 4.9), the drop of piezometric head (e.g. height of the water in an open standpipe at the point considered) will be constant between successive equipotential lines.

$$\Delta h = \frac{h}{N_d}$$

Where h represents the drop in head between the inlet and outlet and N_d is the number of drops. Considering a unit thickness of a flow channel abcd (Fig. 4.8), the hydraulic gradient (i), seepage velocity (v) and amount of flow (q) can be computed as follows:

$$i = \frac{\Delta h}{l_1} = \frac{h}{N_d \times l_1}$$

$$v = k \times i = \frac{kh}{N_d \times l_1}$$

$$\Delta q = (l_2 \times 1) \times \frac{kh}{N_d \times l_1} \text{ as } l_1 = l_2 \text{ then } \Delta q = \frac{kh}{N_d}$$

Thus, the total seepage flow can be computed considering the number of flow channels (N_f) as follows:

$$q = kh \frac{N_f}{N_d} \tag{4.11}$$

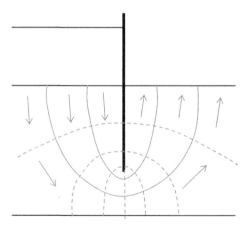

Figure 4.9 Flow net describing the flow under a sheet pile.

Worked-out example 4.1: During a constant head permeameter test, a flow of 173 ml was measured in 5 minutes. A sample of sand having a diameter of 100 mm was tested, and a head difference of 0.61 m was measured between the tapping point (inlet and outlet) located 0.2 m apart. Calculate the permeability of the sand.

Solution: From Darcy's law, we have $q = Aki$. Rearranging for permeability (k), we can obtain

$$k = \frac{q}{Ai}$$

Using the information given:

$$q = \frac{173}{5 \times 60 \times 10^6} = 5.76 \times 10^{-7} \, m^3/s$$

Note conversion to SI units (meters and seconds).

$$A = \frac{0.1^2 \times \pi}{4} = 7.85 \times 10^{-3} \, m^2$$

$$i = \frac{dh}{dl} = \frac{0.61}{0.2} = 3.05 \, m$$

Then

$$k = \frac{5.76 \times 10^{-7}}{7.85 \times 10^{-3} \times 3.05} \quad \text{or} \quad k = 2.4 \times 10^{-5} \, m/s$$

Worked-out example 4.2: A falling head test was conducted on a silty sand having an initial hydraulic head of 900 mm, final head of 400 mm, and 120 seconds was required for the water level in the standpipe to fall. The diameter of the standpipe is 20 mm and the sample diameter is 100 mm and length 180 mm. Calculate the permeability of the soil.

Solution:
From Darcy's law, we have

$$q = Aki$$

In a falling head test (Fig. 4.1b), the hydraulic gradient will change with time due to the variation of the hydraulic head. Thus the flow rate (q) in the previous equation needs to be integrated on the time domain to account for the drop in head in the standpipe having a cross-sectional area of a. In the following equation, the flow rate is represented as a function of the cross-sectional area of the standpipe and the hydraulic head variation (dh) over a given time period (dt), as follows:

$$a\frac{-dh}{dt} = Ak\frac{h}{l}$$

Rearranging, $-a\dfrac{dh}{h} = \dfrac{Ak}{al}dt$ on integration over the initial and final head and the time interval, we obtain

$$-\int_{h_1}^{h_2}\frac{dh}{h} = \frac{Ak}{al}\int_{t_1}^{t_2}dt$$

$$-\ln\frac{h_2}{h_1} = \frac{Ak}{al}(t_2 - t_1) \ \text{ or } \ \ln\frac{h_1}{h_2} = \frac{Ak}{al}(t_2 - t_1)$$

Note: $-\ln(h_2 / h_1) = -\left[\ln(h_1) - \ln(h_2)\right] = \ln(h_2) - \ln(h_1) = \ln(h_1 / h_2)$

Rearranging for permeability, we obtain

$$k = \frac{al\ln(h_1 / h_2)}{A(t_2 - t_1)}$$

Using the information given:

$$a = \frac{0.02^2 \times \pi}{4} = 3.14 \times 10^{-4}\,\text{m}^2$$

$$A = \frac{0.1^2 \times \pi}{4} = 7.85 \times 10^{-3}\,\text{m}^2$$

$h_1 = 0.9$ m

$h_2 = 0.4$ m

$t = 120$ s

$l = 0.18$ m

Then

$$k = \frac{al \ln(h_1 / h_2)}{A(t_2 - t_1)} = \frac{3.14 \times 10^{-4} \times 0.18 \ln(0.9 / 0.4)}{7.85 \times 10^{-3} \times 120} = 4.87 \times 10^{-5} \, \text{m/s}$$

Worked-out example 4.3: A site investigation conducted for a large excavation detected a 5-m-thick layer of clay resting on a 5-m-thick layer of sand, which was underlain by a bed of thick shale. In order to determine the permeability of the sand, a well was driven to the top of the shale and water pumped out at the rate of 10×10^{-3} m³/s. Two observation boreholes were driven through the clay at 15 m and 30 m from the centerline of the well. The water level in these two boreholes was measured, and it was found to be 2 m and 1 m below the ground surface for the borehole located at 15 m and 30 m away, respectively. Calculate the coefficient of permeability of the sand assuming Darcy's law.

Solution: The profile described can be simplified as in Figure 4.10.

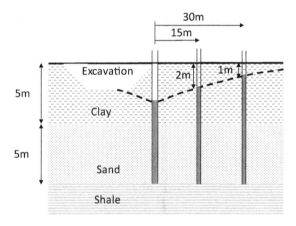

Figure 4.10 Simplified profile for worked-out example 4.3.

From Darcy's law using the confined aquifer equation derived earlier, we have:

$$k = \frac{q \, \ln\left(\dfrac{R_2}{R_1}\right)}{2\pi H \, (H_2 - H_1)}$$

Using the information given, then:

$$q = 10 \times 10^{-3} \, \text{m}^3 / \text{s}$$
$$H = 5 \, \text{m}$$
$$H_1 = (10 - 2) = 8 \, \text{m}$$
$$H_2 = (10 - 1) = 9 \, \text{m}$$
$$R_1 = 15 \, \text{m and } R_2 = 30 \, \text{m}$$

Then:

$$k = \frac{10 \times 10^{-3} \ln \frac{30}{15}}{2\pi \, 5(9-8)} = 2.2 \times 10^{-4} \text{ m/s}$$

Worked-out example 4.4: A sheet pile wall is driven to a depth of 2 m into a layer of sand having a permeability of 6×10^{-3} m/s, which extends to a depth of 3 m below ground level. Below, there is a layer of clay of very low permeability that may be considered impermeable. The wall will impound on one side 1.5 m of water above the level surface of the sand.

a.   Draw the flow net assuming the soil is isotropic.
b.   Determine the approximate seepage flow under the sheet pile wall.

Solution:
a.   The ground profile described may be illustrated as shown in Figure 4.11.

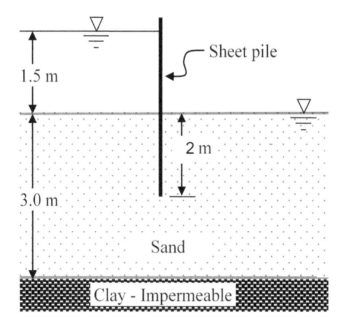

Figure 4.11 Simplified profile for worked-out example 4.4.

The flow lines start and finish at right angles to inlet and outlet surfaces, and equipotential lines start and finish at right angles to impermeable surfaces. A few trial flow lines can be sketched, and these will be smooth curves roughly parallel to the impermeable surfaces as illustrated in Figure 4.12.

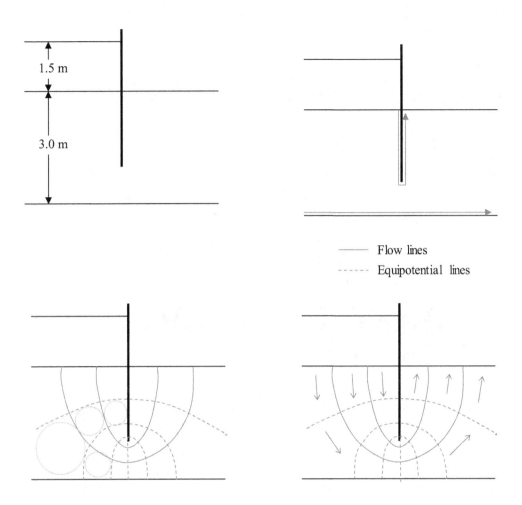

Figure 4.12 Flow net illustration for worked-out example 4.4.

b. Once the flow net is draw, the seepage flow may be calculated considering the geometry of the flow net as follows:

$$q = kh\frac{N_f}{N_d}$$

Where

$$k = 6 \times 10^{-3}\,\text{m/s}$$
$$h = 1.5\ \text{m}$$
$$N_f = 3$$
$$N_d = 8$$

then $q = 6 \times 10^{-3} \times 1.5 \times \dfrac{3}{8} = 3.4 \times 10^{-3} \, \text{m}^3/\text{s}$

Worked-out example 4.5: Figure 4.13 shows the cross-section diagram of a concrete dam spillway built on a pervious soil of 23 m thickness. In flood conditions, it is expected that the impounded water height will reach 19 m and overflow to a height of 6 m on the downstream side.

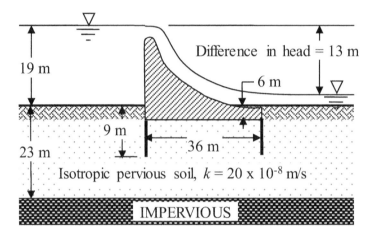

Figure 4.13 Simplified profile of a concrete dam spillway.

a.   Draw the flow net assuming the soil is isotropic.
b.   Calculate the rate of seepage flow under the dam and the uplift pressure on the base.

Solution:
a.   The flow lines start and finish at right angles to inlet and outlet surfaces, and equipotential lines start and finish at right angles to impermeable surfaces. A few trial flow lines can be sketched, and these will be smooth curves roughly parallel to the impermeable surfaces.

The complete flow line may be represented as shown in Figure 4.14.

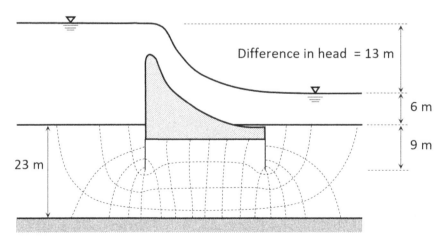

Figure 4.14 Flow net for worked-out example 4.5.

b.   For the flow net represented earlier, we have N_d= 16 and N_f= 4, then

$$q = kh\frac{N_f}{N_d} = 20\times10^{-8}\times13\times\frac{4}{16} = 65\times10^{-8}\,m^3/s \text{ per m run of the dam.}$$

For calculating the uplift pressure, the pressure drop throughout the length of the dam needs to be considered. The pressure drop between two successive equipotential lines can be computed as follows:

$$\Delta h = \frac{h}{N_d} = \frac{13}{16} = 0.81\,m$$

At the upstream base of the dam, the head is 19 m, thus resulting in an uplift pressure of

$$h_w \times \gamma_w = 19\times9.81 = 186.4\,kPa$$

As the flow moves through different equipotentials, there will be a reduction of the pressure directly proportional to the number of drops. The uplift points are shown in Figure 4.15, and associated calculations are tabulated in Table 4.2.

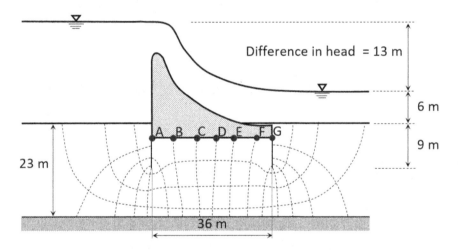

Figure 4.15 Flow net with uplift points.

Table 4.2 Calculation of the points noted in Figure 4.15.

Point	Number of drops	Water head (h_w)	Uplift pressure (kPa) $h_w \times \gamma_w$
A	6	19-6×0.81=14.1 m	138.32
B	7	19-7×0.81=13.3 m	130.5
C	8	19-8×0.81=12.5 m	122.6
D	9	19-9×0.81=11.7 m	114.8
E	10	19-9×0.81=11.7 m	106.9
F	11	19-11×0.81=10.1 m	99.1
G	12	19-12×0.81=9.3 m	91.2

References

Carman, P.C. (1956) *Flow of Gases Through Porous Media*. Butterworths Scientific Publications, London.

Knappett, J. & Craig, R.F. (2012) *Craig's Soil Mechanics*, 8th edition. CRC Press, ISBN:9780415561266, p. 584.

Kozeny, J. (1927) Ueber kapillare Leitung des Wassers im Boden. *Akademie der Wissenschaften in Wien*, 136–271.

Whitlow, R. (1995) *Basic Soil Mechanics*, 3rd edition. Longman Scientific and Technical, London, p. 559.

Chapter 5

Elastic stress and deformation analysis

In the ground, any element of soil will be in equilibrium under normal and shear stresses, acting on three orthogonal axes, x, y and z. These equilibrium stresses will be altered when a load or pressure is applied from the foundation to the underlain soil layer. In most of the geo-technical application, the vertical stress distribution with depth is important to determine the bearing capacity and settlement of foundation. This chapter gives a summary of the methods for the calculation of vertical stress and deformation using theory of elasticity.

5.1 Stresses due to a point load

Boussinesq (1885) determined the solutions to the point load (Q) applied on the surface on a semi-infinite, homogenous, isotropic material with linear stress strain relationship.

The vertical stress (σ_z) at a point A (see Fig. 5.1) is given by:

$$\sigma_z = \frac{Q}{z^2} I_p \tag{5.1}$$

The influence factor I_p depends on the radial distance (r) and depth (z).

$$I_p = \frac{3}{2\pi} \left[\frac{1}{1 + \left(\dfrac{r}{z} \right)^2} \right]^{\frac{5}{2}} \tag{5.2}$$

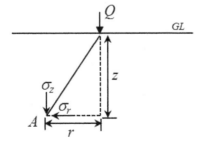

Figure 5.1 Stresses at point A from a point load.

Alternately, Figure 5.2 can be used to calculate I_p.

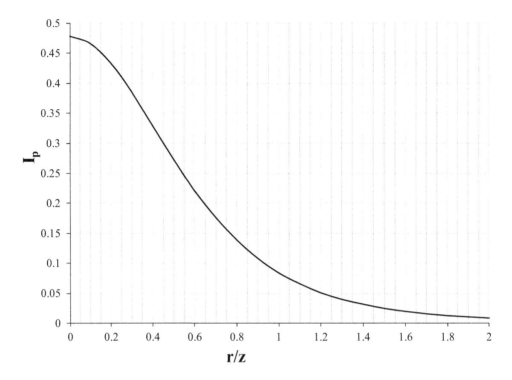

Figure 5.2 Variation of I_p with r/z.

5.2 Stresses due to line load

The vertical stress (σ_z) beneath a line load per unit length (Q/m) on the surface of the ground level (see Fig. 5.3) is given with respect to the depth (z) and horizontal distance (x) is [Fig. 5.3]:

$$\sigma_z = \frac{2Q}{\pi}\left[\frac{z^3}{\left(x^2+z^2\right)^2}\right] \tag{5.3}$$

The lateral stress (σ_x) is given by:

$$\sigma_x = \frac{2Q}{\pi}\left[\frac{x^2 z}{\left(x^2+z^2\right)^2}\right] \tag{5.4}$$

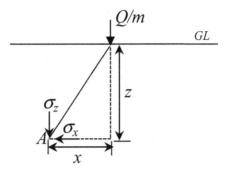

Figure 5.3 Stresses at point A due to line load.

5.3 Stresses due to a uniformly loaded area

Generally, a uniformly applied load can be represented as a large number of point loads. Therefore, integration of the Boussinesq equations will give the stress under a uniform pressure.

5.3.1 Stresses due to uniform strip load

The vertical (σ_z) and horizontal stress (σ_x) at point A due to uniform pressure q in terms of angles α and β (radians) is given by (see Fig. 5.4):

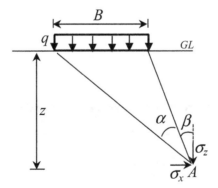

Figure 5.4 Stresses at point A due to uniformly loaded area.

$$\sigma_z = \frac{q}{\pi}\left[\alpha + \sin\alpha \cdot \cos\left(\alpha + 2\beta\right)\right] \tag{5.5}$$

$$\sigma_x = \frac{q}{\pi}\left[\alpha - \sin\alpha \cdot \cos\left(\alpha + 2\beta\right)\right] \tag{5.6}$$

Lines or contours of equal vertical stresses can be plotted using these equations. The zone lying inside the vertical stress contour of $0.2q$ is considered a bulb of pressure.

5.3.2 Stresses due to triangular strip load

The vertical (σ_z) and horizontal stress (σ_x) at point A due to uniform pressure q in terms of angles α and β (radians)and length R_1 and R_2 is given by (see Fig. 5.5):

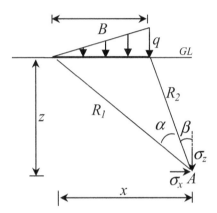

Figure 5.5 Stresses at point A due to uniformly loaded triangular area.

$$\sigma_z = \frac{q}{\pi}\left[\frac{x}{B}\alpha - \frac{1}{2}\sin 2\beta\right] \tag{5.7}$$

$$\sigma_x = \frac{q}{\pi}\left[\frac{x}{B}\alpha - \frac{z}{B}ln\frac{R_1^2}{R_2^2} + \frac{1}{2}\sin 2\beta\right] \tag{5.8}$$

5.3.3 Stresses due to uniformly loaded circular area

The vertical (σ_z) at point A (depth z) due to uniform pressure q is given by:

$$\sigma_z = qI_c$$

The influence factor, I_c, is given in terms of radius (R) and depth (z).

$$I_c = 1 - \left(\frac{1}{1+\left(\frac{R}{z}\right)^2}\right)^{\frac{3}{2}} \tag{5.9}$$

Alternatively, Figure 5.6 can be used to calculate I_c.

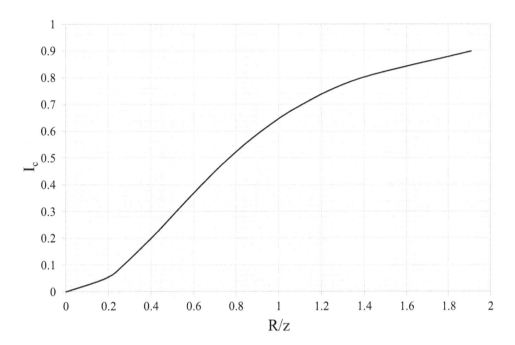

Figure 5.6 Variation of I_c with R/z.

5.3.4 Stresses due to uniformly loaded rectangular area

The vertical stress at a depth z under a corner of a flexible rectangular area of dimensions mz and nz supporting a uniform pressure q is given by:

$$\sigma_z = qI_{qr} \tag{5.10}$$

The influence factor I_{qr} depend on the length (L) and breadth (B) of the loaded area. Figure 5.7 shows the variation of I_{qr} with m and n.

5.4 Vertical stress computation using Newmark Chart

The influence chart prepared by Newmark based on the Boussinesq solution can be used to calculate the vertical stress at any point beneath the loaded area of any shape. In the chart

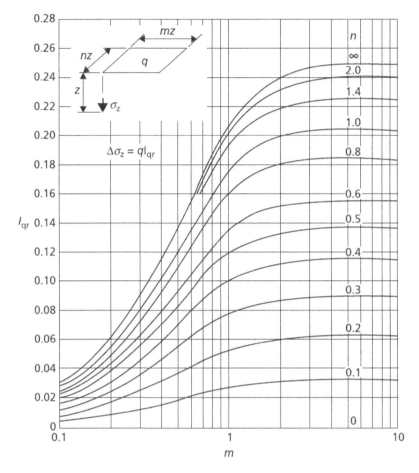

Figure 5.7 Variation of I_{qr} with m and n (after Knappett and Craig, 2012, reproduced with permission from CRC Press).

shown in Figure 5.8, each influence value is 0.005, and each influence block represents a vertical stress of $0.005q$.

1. Draw the loaded area on a tracing paper such that the length of the scale line equals the depth z where the stress is required.
2. Position the drawing on the chart such that the point at which the vertical stress required is at the center of the chart.
3. Count the number of blocks N covered by the loaded area.
4. The vertical stress can then be computed using the equation:

$$\sigma_z = 0.005 \, N \cdot q$$

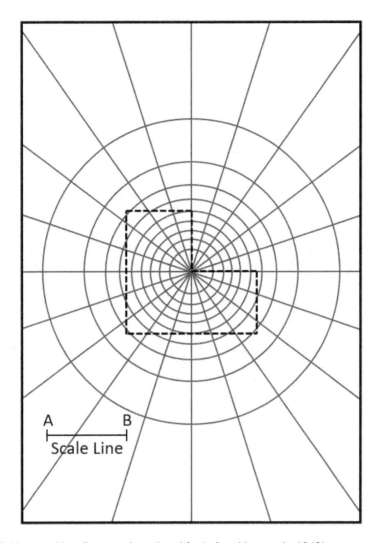

Figure 5.8 Newmark's influence chart (modified after Newmark, 1942).

5.5 Elastic deformations

Elastic theory can be used to predict the vertical deformation beneath an area carrying a uniform pressure q. It is assumed that soil mass is semi-infinite, homogenous and isotropic, with a linear stress–strain relationship. The vertical deformation on the surface of a flexible loaded area can be expressed as:

$$s_i = \frac{q \cdot B}{E}\left(1 - \vartheta^2\right)I_s \tag{5.11}$$

Where B is the least width or diameter of the foundation, E is the modulus of elasticity, ϑ is the Poisson's ratio.

The influence factor, I_s, is the influence factor depends on the shape of the loaded area (see Table 5.1).

Table 5.1 I_s for vertical deformation under flexible area carrying uniform pressure.

Shape of the area	I_s		
	Center	Corner	Average
Square	1.12	0.56	0.95
Rectangle ($L/B = 2$)	1.52	0.76	1.30
Rectangle ($L/B = 5$)	2.10	1.05	1.83
Circle	1.00	0.64	0.85

Generally, the soil deposit will be of limited thickness and will be underlain by very stiff or hard stratum. For these cases, the average vertical settlement beneath a flexible area carrying a uniform pressure q can be computed using the Giroud (1972) and Burland (1970) equations:

$$s_i = \mu_0 \mu_1 \frac{q \cdot B}{E} \tag{5.12}$$

Where μ_0 and μ_1 are coefficients depends on the depth and breadth of the foundation and also depends on the layer thickness; B is the least width or diameter of the foundation, E is the modulus of elasticity. Figure 5.9 gives the value of μ_0 and μ_1 for Poisson's ratio (ϑ) of 0.5.

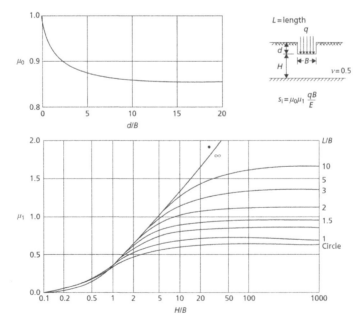

Figure 5.9 Coefficient of vertical displacement μ_0 and μ_1 for Poisson's ratio ($v = 0.5$) (after Knappett and Craig, 2012, reproduced with permission from CRC Press).

Eq.5.12 provides an estimate of immediate settlement of foundation resting on saturated clays. Immediate settlement occurs mainly under undrained conditions; therefore, the appropriate value of ϑ is 0.5. Appropriate field or laboratory testing can be used to determine the modulus of elasticity (E) of the soil (e.g. pressuremeter test, triaxial test).

Worked-out example 5.1: A concentrated load of 850 kN is applied at the ground surface. Calculate the vertical stress (a) at a depth 4 m directly below the load (b) at a radial distance of 2 m for a depth of 6 m from the load.

Solution:
a.   At a depth 4 m directly below the load ($r = 0$):

$$Q = 850 \text{ kN}$$

$$\frac{r}{z} = 0$$

From Figure 5.2, $I_p = 0.48$

$$\sigma_z = \frac{Q}{z^2} I_p = \frac{850}{4^2} \times 0.48 = 25.5 \text{ kPa}$$

b. $r = 2$ m and $z = 6$ m

From Figure 5.2, $I_p = 0.36$.

$$\sigma_z = \frac{Q}{z^2} I_p = \frac{850}{6^2} \times 0.36 = 8.5 \text{ kPa}$$

Worked-out example 5.2: Determine the vertical stress under the point A below a rectangular area 4 m × 8 m (see Fig. 5.10) at a depth of 6 m from the ground level. This area is subjected to a uniform pressure of 200 kN/m².

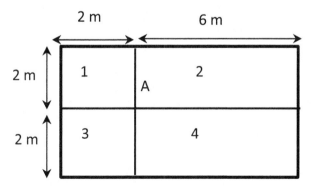

Figure 5.10 Diagram for worked-out example 5.2.

Solution:

$$\sigma_z = qI_r$$

Area 1: 2 m × 2 m

$$mz = 2$$

$$m = \frac{2}{6} = 0.33$$

$$nz = 2;$$

$$n = \frac{2}{6} = 0.33$$

From Figure 5.7, $I_r = 0.045$.

Area 2: 6 m × 2 m

$$mz = 2$$

$$m = \frac{2}{6} = 0.33$$

$$nz = 6;$$

$$n = \frac{6}{6} = 1$$

From Figure 5.7, $I_r = 0.085$.

Area 3: 2 m × 2 m

$$mz = 2$$

$$m = \frac{2}{6} = 0.33$$

$$nz = 2;$$

$$n = \frac{2}{6} = 0.33$$

From Figure 5.7, $I_r = 0.045$.

Area 4: 6 m × 2 m

$$mz = 2$$

$$m = \frac{2}{6} = 0.33$$

$$nz = 6$$

$$n = \frac{6}{6} = 1$$

From Figure 5.7, $I_r = 0.085$.

$$\sigma_z = q\left(I_{r_1} + I_{r_2} + I_{r_3} + I_{r_4}\right) = 200 \times (2 \times 0.045 + 2 \times 0.085) = 52 \text{ kPa}$$

Worked-out example 5.3: A 3 m wide strip footing with a uniform pressure of 150 kPa is founded on the surface of granular soil deposit. The water table is at the ground surface. Determine the effective vertical and lateral stresses at a point 4 m below the center of the footing before and after the application of the load. The saturated unit weight of the soil is 19.5 kN/m³, and $K_0 = 0.5$.

Solution:
  After loading:
  At a point 4 m below the center of the footing:

$$\alpha = 2\tan^{-1}\left(\frac{1.5}{4}\right) = 0.72 \text{ radians}$$

$$\sin\alpha = 0.657$$

$$\beta = -0.36$$

$$\Delta\sigma_z = \frac{q}{\pi}\left[\alpha + \sin\alpha \cdot \cos(\alpha + 2\beta)\right] = \frac{150}{\pi}\left[0.72 + 0.657\right] = 66 \text{ kPa}$$

$$\Delta\sigma_x = \frac{q}{\pi}\left[\alpha - \sin\alpha \cdot \cos(\alpha + 2\beta)\right] = \frac{150}{\pi}\left[0.72 - 0.657\right] = 3 \text{ kPa}$$

Before loading:

$$\sigma_z' = 4 \times 9.69 = 38.76 \text{ kPa}$$

$$\sigma_x' = 0.5 \times 38.76 = 19.38 \text{ kPa}$$

Therefore,

$$\sigma_z' = 38.76 + 66 = 104.76 \text{ kPa}$$

$$\sigma_x' = 19.38 + 3 = 22.38 \text{ kPa}$$

Worked-out example 5.4: A pressure of 40 kPa is uniformly distributed over the area as shown in Figure 5.11. Determine the vertical stress at 3 m below point A using influence factors and Newmark's chart.

Solution:

Method 1: using influence factors

By dividing the footing into three regions 1, 2 and 3 (Fig. 5.11) with area 2 m × 2 m,

$$mz = 2$$

$$m = \frac{2}{3} = 0.67$$

$$nz = 2;$$

$$n = \frac{2}{3} = 0.67$$

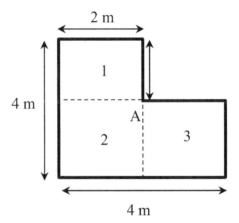

Figure 5.11 Diagram for worked-out example 5.4.

From Figure 5.7, $I_r = 0.13$.

$$\sigma_z = q\left(I_{r_1} + I_{r_2} + I_{r_3}\right) = 40 \times (3 \times 0.13) = 16 \text{ kPa}$$

Method 2: using Newmark's Chart

As presented in Figure 5.8, draw Figure 5.11 on tracing paper to a scale line equals the depth (say 3 m). Position the diagram such way that the point A coincides with the center of the chart. Count the number of influence area covered by the diagram, $N \cong 85$. Then,

$$\sigma_z = 0.005 \times N \times q = 0.005 \times 85 \times 40 = 17 \text{ kPa}$$

Worked-out example 5.5: A square foundation having dimensions 5 m ×5 m exerts a uniform pressure of 150 kPa on a semi-infinite saturated soil ($E = 40$ MPa). Determine the value of immediate settlement at the center of the foundation (assume $\vartheta = 0.5$).

Solution:

$$s_i = \frac{q \cdot B}{E}\left(1 - \vartheta^2\right) I_s$$

From Table 5.1, $I_s = 1.12$.

$$s_i = \frac{150 \times 5}{40 \times 10^3}\left(1 - 0.5^2\right)1.12 = 15.75 \text{ mm}$$

Worked-out example 5.6: A rectangular foundation 3 m ×1.5 mis located at a depth of 2 m in a layer of clay of 6 m deep and underlain by a hard stratum. It carries a uniform pressure of 100 kPa. The clay is saturated, and the value of E_u is 40 MN/m². Determine the average immediate settlement under the foundation.

Solution:

$$s_i = \mu_0\mu_1\frac{q \cdot B}{E}$$

$$\frac{d}{B} = \frac{2}{1.5} = 1.33 \text{ From Figure 5.9, } \mu_0 = 0.92$$

$$\frac{H}{B} = \frac{4}{1.5} = 2.67 \text{ From Figure 5.9, } \mu_1 = 0.60$$

$$\frac{L}{B} = \frac{3}{1.5} = 2$$

$$s_i = \mu_0\mu_1\frac{q \cdot B}{E} = 0.92 \times 0.6 \times \frac{100 \times 1.5}{40 \times 10^3} = 2.1 \text{ mm}$$

Worked-out example 5.7: A flexible foundation 2.5 m square is to carry a uniformly distributed load of 250 kN. The foundation is founded at a depth of 3 m below the surface of the clay. The value of E obtained from the triaxial compression test is 18 MPa. The clay stratum is 10 m deep and underlain by a thick dense sandy gravel (assume clay is saturated and v = 0.5).

Solution:

$$\frac{d}{B} = \frac{3}{2.5} = 1.2 \text{ From Figure 5.9, } \mu_0 = 0.91$$

$$\frac{H}{B} = \frac{7}{2.5} = 2.8 \text{ From Figure 5.9, } \mu_1 = 0.55$$

$$\frac{L}{B} = \frac{2.5}{2.5} = 1$$

$$s_i = \mu_0\mu_1\frac{q \cdot B}{E} = 0.91 \times 0.55 \times \frac{250 \times 2.5}{18 \times 10^3} = 17.37 \text{ mm}$$

References

Boussinesq, M.J. (1885) *Applications des potentiels à l'étude de l'équilibre et mouvement des solides elastiques*. Gauthier-Villard, Paris.

Burland, J.B. (1970) Discussion. *Proceedings of Conference on In-Situ Investigations in Soils and Rocks*, British Geotechnical Society, London, p. 61.

Giroud, J.P. (1972) Settlement of rectangular foundation on soil layer. *Journal of the ASCE*, 98(SM1), 149–154.

Knappett, J. & Craig, R.F. (2012) *Craig's Soil Mechanics*, 8th edition. CRC Press, ISBN:9780415561266, p. 584.

Newmark, N.M. (1942) Influence charts for computation of stresses in elastic foundations. *University of Illinois Engineering Experiment Station Bulletin*, (338).

Chapter 6

Consolidation settlement and analysis

When a saturated soil layer is subjected to an increase in load from a foundation, the soil layer experiences a corresponding increase in pore water pressure. The dissipation rate of excess pore water pressure depends on the permeability of the soils. The dissipation of pore water pressure can be rapid for coarse-grained soil (gravel and sand) and slow for fine-grained soils (clay and silts). The dissipation of pore pressure and associated drainage of water can lead to a compression of the soil layer. This time-dependent process of volume reduction due to the expulsion of water from the voids is defined as consolidation. The process describing the dissipation of excess pore water pressure and associated settlement can be captured using consolidation theory.

The one-dimensional consolidation theory proposed by Terzaghi (1925) is widely used for estimating the consolidation settlements of soils.

6.1 Consolidation settlement

One-dimensional consolidation tests (oedometer test) are generally performed to determine stress–strain behavior in confined compression (no lateral deformation) and to obtain associated consolidation parameters to estimate the consolidation settlement and time.

The volume change (ΔV) can be determined using the change in height (ΔH) or change in void ratio (Δe) (Fig. 6.1),

$$\frac{\Delta V}{V} = \frac{\Delta H}{H} = \frac{\Delta e}{1+e_0} \tag{6.1}$$

where e_0 is the initial void ratio of the soil sample.

The consolidation settlement (s_c) can be calculated using the change in void ratio as

$$s_c = m_v \Delta \sigma' H \tag{6.2}$$

where m_v is the coefficient of volume compressibility, $\Delta \sigma'$ is the change in the effective stress and H is the initial soil layer thickness.

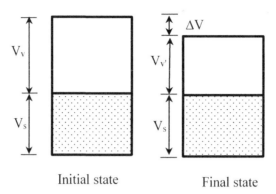

Figure 6.1 Soil sample during consolidation.

6.2 Preconsolidation pressure

Preconsolidation pressure is defined as the maximum pressure the soil has experienced in the past. A careful evaluation of preconsolidation pressure is important for the computation of consolidation settlement. An estimate on the preconsolidation pressure can be obtained from the graphical procedure proposed by Casagrande (1936). The pressure can also be used as a limiting pressure for controlling long term settlements in overconsolidated soils.

6.3 Overconsolidation ratio

The overconsolidation ratio (OCR) is defined as the ratio of the past maximum pressure (σ'_0) on the soil to the current pressure (σ'_1). This ratio also provides an assessment on the degree of overconsolidation.

$$OCR = \sigma'_0 / \sigma'_1 \tag{6.3}$$

Normally consolidated soil: If the pressure the soils was exposed to in the past is equal to the current pressure $\sigma'_0 = (\sigma'_1)$, then $OCR = 1$. The consolidation settlement can be determined from Eq. 6.4. The value of compression index (c_c) can be determined from the e-log σ'_v data obtained from the laboratory experiments (see Fig. 6.2).

$$s_c = \frac{c_c}{1+e_0} H log \frac{\sigma'_1}{\sigma'_0} \tag{6.4}$$

Skempton (1944) suggested an empirical relationship between compression index (C_c) and liquid limit (w_L) :

For undisturbed clays $c_c = 0.009(w_L - 10)$ \hfill (6.5)

For remoulded clays $c_c = 0.007(w_L - 10)$ \hfill (6.6)

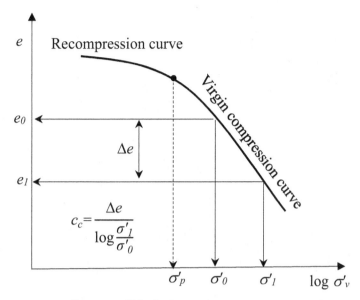

Figure 6.2 Plot of $e - log\ \sigma_v$ for normally consolidated soil.

Overconsolidated soil: If the past pressure is greater than current pressure $(\sigma'_0 > \sigma'_1)$, then $OCR > 1$. The consolidation settlement in this state can be determined from Eq. 6.7. The value of recompression index (C_r) can be determined from the e−log σ'_v data obtained from the laboratory experiments (see Fig. 6.3):

$$s_c = \frac{c_r}{1+e_0} Hlog \frac{\sigma'_1}{\sigma'_0} \tag{6.7}$$

6.4 Rate of consolidation settlement

The classical theory proposed by Terzaghi (1925) can be effectively used for estimating the rate of consolidation. Many key assumptions have been considered for the development of the one-dimensional consolidation theory.

The differential equation for consolidation relating to excess pore water pressure (u), depth (z) and time (t) can be written as:

$$\frac{\partial u}{\partial t} = c_v \frac{\partial^2 u}{\partial z^2} \tag{6.8}$$

where c_v is the coefficient of consolidation given by the ratio $k/(\gamma_w m_v)$ and k is the permeability of soil, m_v is the coefficient of volume compressibility and γ_w is the unit weight of water.

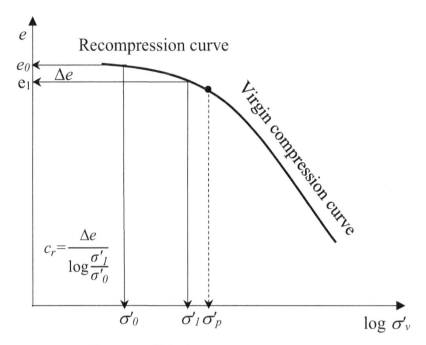

- Preconsolidation pressure

c_r Reompression index

Figure 6.3 Plot of $e-log\ \sigma'_v$ for overconsolidated soil.

The relationship between average degree of consolidation (U) and dimensionless time factor (T_v) can be shown as

$$U = \frac{S_t}{S} = 1 - \sum_{m=0}^{m=\infty} \frac{2}{M^2} e^{-M^2 T_v} \tag{6.9}$$

Where U is average degree of consolidation, S_t is settlement at any time t, S is ultimate settlement of the layer from primary consolidation, $M = (\pi/2)\,(2\,m + 1)$, and m is an integer.

$$T_v = \frac{c_v t}{d^2} \tag{6.10}$$

Where d is the drainage depth. It represents the maximum distance water molecules have to travel to escape from the soil layer. For an open layer (drainage occurs at the top and bottom of the soil layer), d is half the thickness of the soil layer, and for a half-closed layer (drainage occurs at either the top or bottom of the soil layer), d is the total thickness of the soil layer.

Approximate relationship between average degree of consolidation (U) and time factor (T_v; Fig. 6.4) can be obtained from:

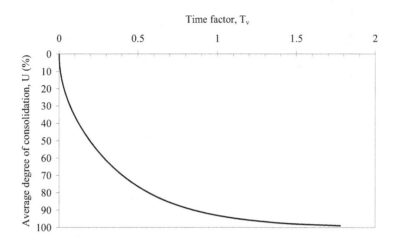

Figure 6.4 Relationship between average degree of consolidation and time factor.

For $U = 0$ *to* 60% $Tv = \dfrac{\pi}{4}\left(\dfrac{U\%}{100}\right)^2$ (6.11a)

For $U > 60\%$ $T_v = 1.781 - 0.933 \log(100 - U\%)$ (6.11b)

The coefficient of consolidation c_v is an important parameter for estimating the rate of settlement. It can be determined from the consolidation test data using the curve-fitting methods proposed by Taylor (1942) or Casagrande and Fadum (1940).

6.5 Secondary compression of clays

Compression in soils does not cease after the complete dissipation of excess pore water pressure developed during loading (i.e. at the end of primary compression). However, it continues at a decreased rate at constant effective stress due to the soil particle readjustment. This soil readjustment is mainly due to the viscous flow of adsorbed water surrounding the soil particle. The compression of soil at this stage of the consolidation process is referred to as secondary compression or creep (Fig. 6.5). For high-plastic clays and organic clays (e.g. peat), secondary compression will be significant compared to the primary compression.

The coefficient of secondary compression (C_α) is defined as the ratio of the change in void ratio (Δe) per log cycle of time (log t) scale.

$$C_\alpha = \frac{\Delta e}{\Delta \log t}$$ (6.12)

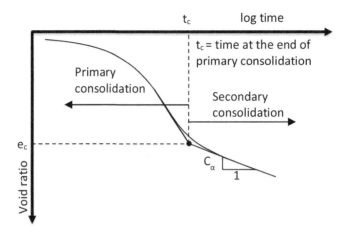

Figure 6.5 Void ratio versus log time plot – secondary compression.

Mesri and Godlewski (1977) have suggested the ratio of $\dfrac{C_\alpha}{C_c}$ in the range of 0.025–0.10 for natural soils.

The amount of secondary compression can be calculated from:

$$S_s = C_{\alpha e} \, H_c \, log\left(\frac{t}{t_c}\right) \tag{6.13}$$

Where $C_{\alpha e}$ is called modified secondary compression index $= \dfrac{C_\alpha}{1+e_c}$; H_c is the thickness of the clay layer at the end of primary consolidation; e_c is the void ratio at the end of primary consolidation (e_0 can also be used without any loss of accuracy); t_c is the time at the end of primary consolidation. Figure 6.6 shows the modified secondary compression index for natural deposits.

Worked-out example 6.1: A normally consolidated clay stratum 10 m thick is located at a depth 15 m below ground level. The soil is sand above the clay stratum. The water table is located 10 m below the ground level. The submerged unit weight of the sand (γ_{sub}) is 12 kN/m³ and the unit weight is 18 kN/m³ above the water table. The increase in pressure at the center of the clay stratum is 150 kN/m² due to the weight of a building. Estimate the expected ultimate settlement of the structure due to primary consolidation.

Properties of the clay layer:
  Natural moisture content: 45%
  Liquid limit: 52%
  Specific gravity: 2.77

Solution:
  The ground profile is shown in Figure 6.7.

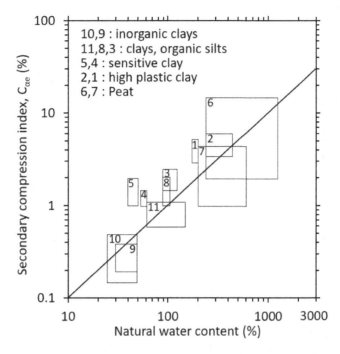

Figure 6.6 Modified secondary compression index, C_{ae} (modified after Mesri, 1973).

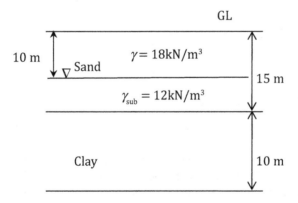

Figure 6.7 Profile for worked-out example 6.1.

Settlement for normally consolidated soil:

$$S_c = \frac{c_C}{1+e_0} H \log\left(\frac{\sigma_1'}{\sigma_0'}\right)$$

$c_C = 0.009\left(w_L - 10\right)$ from Eq. 6.5

$$c_C = 0.009(52-10)$$

$$c_C = 0.38$$

Clay

$$e = wG = 0.45 \times 2.77 = 1.25 \quad \gamma_{sat} = \left(\frac{G+e}{1+e}\right)\gamma_W$$

$$\gamma_{Sat} = \left(\frac{2.77+1.25}{1+1.25}\right)9.81$$

$$\gamma_{Sat} = 17.52 \text{ kN/m}^3$$

At $z = 20$ m from GL:

Initial condition:

$$\sigma_0' = (10 \times 18) + (12 \times 5) + ((17.52 - 9.81) \times 5)$$

$$\sigma_0' = 278.55 \text{ kPa}$$

Considering the pressure from the weight of the building:

$$\sigma_0' = (10 \times 18) + (12 \times 5) + ((17.52 - 9.81) \times 5) + 150$$

$$\sigma_1' = 240 + 188.55$$

$$\sigma_1' = 428.55 \text{ kPa}$$

$$S_c = \frac{0.38}{1+1.25} \times 10 \times \log\left(\frac{428.55}{278.55}\right)$$

$$S_c = 0.32 \text{ m}$$

Worked-out example 6.2: The results obtained from a consolidation testing of a fully saturated clay sample are given in Table 6.1. Determine the void ratio corresponding to each pressure increment and plot void ratio versus log σ_v'. Calculate compression index c_c and pre-consolidation pressure (σ_p).

Initial thickness (H_0) = 20 mm
Final moisture content = 26%
Specific gravity of clay particles = 2.76

Table 6.1 Data for worked-out example 6.2.

Applied pressure (kPa)	Thickness of specimen (mm)
0	20.00
25	19.79
50	19.73
100	19.5
200	19.26
400	18.72
800	18.2
0	18.6

Solution:

Compute final void ratio:

$$e_1 = W_1 G$$

$$e_1 = 0.26 \times 2.76 = 0.72$$

Void ratio at the start of the test:

$$e_0 = e_1 + \Delta e$$

$$\Delta e = \frac{\Delta H}{H_0}(1 + e_0)$$

$$\Delta e = \frac{1.4}{20}(1 + e_1 + \Delta e)$$

$$\Delta e = \frac{1.4}{20}(1 + 0.72 + \Delta e)$$

$$\Delta e = 0.13$$

$$e_0 = e_1 + \Delta e = 0.72 + 0.13 = 0.85$$

$$\frac{\Delta e}{\Delta H} = \frac{1.85}{20} = 0.093$$

A detailed solution for all pressure increment is presented in Table 6.2.
Compression index (C_c) in the stress range on the linear part of the *e*-log σ_v. plot:

Compression index, $c_c = \dfrac{\Delta e}{log \dfrac{\sigma'_1}{\sigma'_0}} = 0.166$

Using the graphical procedure proposed by Casagrande (1936); refer to Section 6.2:
Preconsolidation pressure (σ_p) = 167 kPa

Table 6.2 Solution of worked-out example 6.2.

Pressure (kPa)	Δσ	H (mm)	ΔH (mm)	Δe	e
0		20			0.85
	25		−0.21	−0.02	
25		19.79			0.83
	25		−0.06	−0.006	
50		19.73			0.82
	50		−0.23	−0.021	
100		19.5			0.80
	100		−0.24	−0.022	
200		19.26			0.78
	200		−0.54	−0.050	
400		18.72			0.73
	400		−0.52	−0.048	
800		18.2			0.68
			0.4	0.037	
0		18.6			0.72

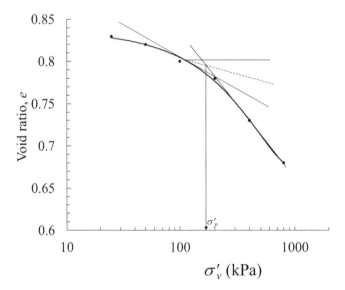

Figure 6.8 Plot of void ratio versus log σ_v.

Worked-out example 6.3: A highway embankment is 6 m above the existing ground level. The ground layer is a soft clay deposit of 10 m thick. The clay soil has an $m_v = 0.62$ m²/MN and $c_v = 9.5$ m²/year. If the tolerable settlement is 50 mm for the road pavement, then estimate time for placing the pavement. (Assume bulk unit weight of embankment = 21 kN/m³.) Assume sand layer below the soft clay deposit.

Solution: The ground profile is illustrated in Figure 6.9. The amount of consolidation settlement can be computed using Eq. 6.2:

$$S_C = m_V \, H \, \Delta\sigma$$

$$\Delta\sigma = 6 \times 21$$

$$\Delta\sigma = 126 \text{ kPa}$$

$$S_C = \frac{0.62}{10^6} \times 10 \times 126 \times 10^3$$

$$S_C = 0.78 \, \text{m}$$

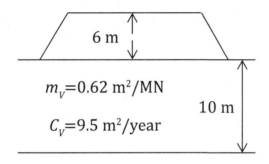

$m_V = 0.62 \text{ m}^2/\text{MN}$

$C_V = 9.5 \text{ m}^2/\text{year}$

6 m

10 m

Figure 6.9 Profile for worked-out example 6.3.

To ensure 6-m-high embankment, the amount of fill should be $(6 + S_c)$ m.

$$S_C = \frac{0.62}{10^6} \Big[(6 + S_C) \times 10^3 \times 21 \Big] \times 10$$

$$S_C = \frac{10}{10^3} \big[78.12 + 13.02 \, S_C \big]$$

$$S_C = 0.7812 + 0.1302 \, S_C$$

$$0.87 \, S_C = 0.7812$$

$$S_C = 0.90 \, \text{m}$$

Degree of consolidation during the placement of the pavement should be:

$$U = \frac{900 - 50}{900} = 0.94$$

From Eq. 6.11:

$$T_V = 1.781 - 0.933 \log(100 - U\%)$$

$$T_V = 1.05$$

$$t = \frac{Td^2}{C_V}$$

$$t = \frac{1.05 \times 5^2}{9.5}$$

$$t = 2.76 \text{ Years}$$

Worked-out example 6.4: A pressure of 50 kPa was applied to a clay layer of 10 m thick. The water table is at the ground level. Determine the excess pore water pressure at the middle of the clay layer after 6 months assuming the load was applied instantaneously. The c_v value of clay is 12 m²/yr. Assume impermeable stratum below the clay layer and bulk unit weight of 21 kN/m³.

Solution:
Hydrostatic pore water pressure before placing the fill = 5 × 9.81= 49.05 kPa
Effective stress (initial) = 21 × 5–5 × 9.81 = 55.95 kPa
Excess pore pressure at the middle of clay layers = 50 kPa.
After 6 months,

$$T_V = \frac{12 \times 6}{10^2 \times 12}$$

$$T_V = 0.06$$

Using Eq.6.10: $T_V = \frac{\pi}{4} U^2$

$$U = 0.28$$

Excess pore water pressure dissipated = 50 × 0.28
= 14 kPa
Excess pore water pressure remaining = 50 × 0.72
= 36 kPa
Pore pressure in the middle of the clay at this time = 49.05 + 36
= 85.05 kPa

Worked-out example 6.5: In a consolidation test, a fully saturated clay specimen of 20 mm thick reaches 50% degree of consolidation in 18 min. How long would it take an 8-m thick layer of clay to reach the same degree of consolidation? Assume the same stress and boundary condition.

Solution: Consolidation tests are generally conducted with two-way drainage and thus drainage path in the consolidation test:

$$d = \frac{20 \times 10^{-3}}{2}$$

$$d = 10 \times 10^{-3} \text{ m}$$

$$c_V = \frac{T_v d^2}{t}$$

$$\frac{T_{lab} d_{lab}^2}{t_{lab}} = \frac{T_f d_f}{t_f}$$

$$\frac{\left(10 \times 10^{-3}\right)^2}{18} = \frac{(4)^2}{t_f}$$

$$t_f = \frac{(16 \times 18)}{\left(10 \times 10^{-3}\right)^2}$$

$$t_f = 2880000 \text{ minutes}$$

$$t_f = 5.5 \text{ years}$$

Worked-out example 6.6: Determine the magnitude of secondary compression of an embankment constructed on soft high plastic clay for a period of 10 years. The secondary compression index $(C_{\alpha e}) = 1.6\%$. The high-plastic clay layer is 9 m thick. Assume t_c is 1 year and negligible primary compression settlement.

Solution:

$H' = 9.0$ m assuming negligible primary consolidation

$$s_s = C_{\alpha e} H_c \log\left(\frac{t}{t_c}\right)$$

$$s_s = 0.016 \times 9 \times \log\left(\frac{10}{1}\right)$$

$$s_s = 144 \text{ mm}$$

Worked-out example 6.7: An embankment is planned to be constructed on organic clay having a thickness of 5 m. The initial void ratio and compression index of the clay is 0.75 and 0.6, respectively. Estimated time for completing the primary consolidation is 1.3 years. Calculate the total settlement of the clay 7 years after the completion of primary consolidation. The average effective pressure on the clay layer $(\sigma'_0) = 100$ kPa and $\Delta\sigma' = 25$ kPa. Assume secondary compression index $(C_{\alpha e}) = 1.6\%$.

Solution:

The void ratio at the end of primary compression e_c:

$$e_c = e_0 - \Delta e_c$$

$$\Delta e' = C_c log \left[\frac{\sigma_o' + \Delta \sigma'}{\sigma_o'} \right] = 0.6 \times log \left[\frac{100 + 25}{100} \right] = 0.058$$

Therefore,

$$e_c = e_0 - \Delta e_c = 0.75 - 0.058 = 0.7$$

The amount of primary compression:

$$s_c = \frac{\Delta e \, H}{1 + e_0} = \frac{0.058 \times 5}{1 + 0.75} = 0.17 \text{ m}$$

$$H_c = H - s_c = 5 - 0.17 = 4.83 \text{ m}$$

The amount of secondary compression:

$$s_s = C_{\alpha e} \, H_c \, log \left(\frac{t}{t_c} \right) = 0.016 \times 4.83 \times log \left(\frac{7}{1} \right) = 0.063 \text{ m}$$

The total consolidation settlement = 0.17 + 0.056 = 0.226 m.

References

Casagrande, A. (1936) Determination of the preconsolidation load and its practical significance. *Proceedings, 1st International Conference on Soil Mechanics and Foundation Engineering*, Cambridge, MA, vol. 3, pp. 60–64.

Casagrande, A. & Fadum, R.E. (1940) Notes on soil testing for engineering purposes. *Harvard University Graduate School of Engineering Publication*, (8).

Mesri, G. (1973) Coefficient of secondary compression. *Journal of Geotechnical Engineering, ASCE*, 99(1), 123–137.

Mesri, G. & Godlewski, P.M. (1977) Time-and stress-compressibility interrelationship. *Journal of Geotechnical Engineering, ASCE*, 103(5), 417–430.

Skempton, A.W. (1944) Notes on the compressibility of clays. *Quarterly Journal of the Geological Society*, 100(1–4), 119–135.

Taylor, D.W. (1942) *Research on Consolidation of Clays*, Serial No. 82. Department of Civil and Sanitary Engineering, Massachusetts Institute of Technology, Cambridge, MA.

Terzaghi, K. (1925) *Erdbaumechanik auf Bodenphysikalischer Grundlage*. Deuticke, Vienna.

Chapter 7

Shear strength, stress paths and failure

This chapter focuses on the evaluation of shear strength parameters from results of tests and their relationship between the shear stress parameter for different drainage conditions.

7.1 Mohr-Coulomb model

The strength of a soil depends on its resistance to shearing stresses that is provided by the friction at the particle contacts and cohesive resistance for clay soil under certain conditions. In general, the state of stress in an element of soil can be defined in terms of normal and shear stresses applied at the boundaries, which can be represented in a Mohr circle (Fig. 7.1). The circle can be used to capture all possible stress states in the soil element. At point P where the circle intersects the failure envelope indicates that the soil has reached failure. Typically the failure envelope can be represented by a straight line, as follows:

$$\tau = c + \sigma \tan \phi \tag{7.1}$$

where c and ϕ are the shear strength parameters, i.e. cohesion intercept and angle of shearing resistance, respectively. As illustrated in Figure 7.1, there is a relationship between the plane defined by $2\theta = 90° + \phi$ in the diagram and the plane θ in the soil element.

The coordinates of point P or the stresses on Failure Plane AB can be defined using trigonometry as follows:

$$\sigma_n = \frac{\sigma_1 + \sigma_3}{2} + \frac{\sigma_1 - \sigma_3}{2} \cos 2\theta \tag{7.2}$$

$$\tau_s = \frac{\sigma_1 - \sigma_3}{2} \sin 2\theta \tag{7.3}$$

and

$$2\theta = 90° + \phi \tag{7.4}$$

$$\theta = 45° + \phi / 2 \tag{7.5}$$

Additional relationships can be established based on Figure 7.1 as follows:

$$\sin \phi = \frac{CP}{CN + NM} = \frac{(\sigma_1 - \sigma_3)/2}{(\sigma_1 + \sigma_3)/2 + c \cot \phi} \tag{7.6}$$

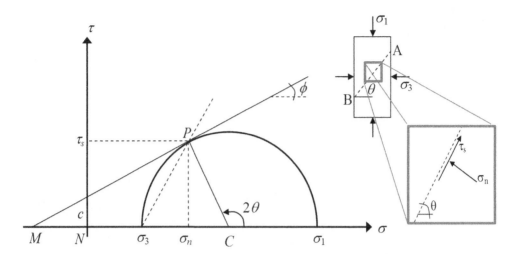

Figure 7.1 Mohr circle representation.

$$\sigma_1 - \sigma_3 = \left(\sigma_1 + \sigma_3\right)\sin\phi + 2c\cos\phi \tag{7.7}$$

Eq. 7.7 can be further rearranged as follows:

$$\sigma_1 = \sigma_3\left(\frac{1+\sin\phi}{1-\sin\phi}\right) + 2c\frac{\cos\phi}{1-\sin\phi} \tag{7.8}$$

Eq. 7.8 is often referred to as the Mohr-Coulomb failure criterion, and it defines the relationship between the principal stresses at failure for given shear strength parameters (e.g. c and ϕ).

In addition, given the stress state, the Mohr circles for total and effective stress have the same diameter but their centers are separated by the corresponding pore water pressure (u) magnitude. The effective principal stress can be calculated in accordance with effective stress relationship ($\sigma' = \sigma - u$).

7.2 Laboratory tests for determining shear strength parameters

There are two main types of drainage conditions that can be adopted to determine the shear strength parameters, i.e. drained and undrained conditions. When using the results of these laboratory tests to analyze the behavior of geomaterials applied in construction, the main consideration is the ratio between the rate at which the changes in total stress and dissipation of pore water pressure occur. Typically, undrained conditions represent short-term analysis (e.g. days and a few weeks during construction), whereas drained conditions correspond to long-term (e.g. years and decades, during service life) analysis.

7.2.1 Direct shear tests

In this test, a number of soil specimens (minimum 3) is tested, each under different vertical stress (σ_v), and the value of the shear stress at failure (τ_f) is plotted against the normal effect stress (σ'_f) as illustrated in Figure 7.2. As there is no control on the drainage conditions, the specimens must be sheared using a small enough shear rate to ensure fully drained conditions are achieved. Only an approximation of the state of pure shear is produced in the specimens during these tests, and as the cross-sectional area varies during shearing, the vertical stress does not remain constant.

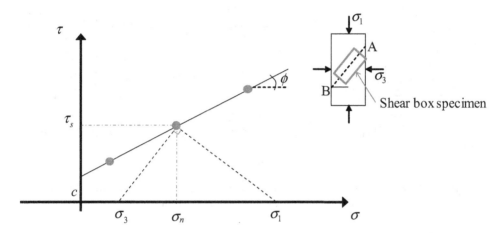

Figure 7.2 Representation of the direct shear tests results in the τ-σ plane.

7.2.2 Triaxial test

In the triaxial test, the drainage can be controlled, and if undrained conditions are adopted, the test is accompanied by pore water pressure measurement. The specimens can be tested for different all around confining pressures, and shearing is induced by gradually increasing the axial load (Fig. 7.3). Both triaxial compression and extension loading modes are possible, albeit compression is more widely tested.

Triaxial test data may be presented in terms of Mohr circles at failure but can also be represented in terms of stress invariants, such that a given set of effective stress conditions can be represented by a single point instead of a circle. For 3D cases, the commonly adopted stress invariants are mean stress invariant or p and deviatoric stress invariant or q, as follows:

$$p = \frac{\sigma_1 + \sigma_2 + \sigma_3}{3} \tag{7.9}$$

$$q = \frac{1}{\sqrt{2}}\left[\left(\sigma_1 - \sigma_2\right)^2 + \left(\sigma_2 - \sigma_3\right)^2 + \left(\sigma_3 - \sigma_1\right)^2\right]^{\frac{1}{2}} \tag{7.10}$$

The results of triaxial tests can be plotted either in the τ-σ, where the confining pressures and shearing load at failure are represented by the equivalent principal stress σ_3 and σ_1 Mohr circles, or in terms of p and q (Fig. 7.4).

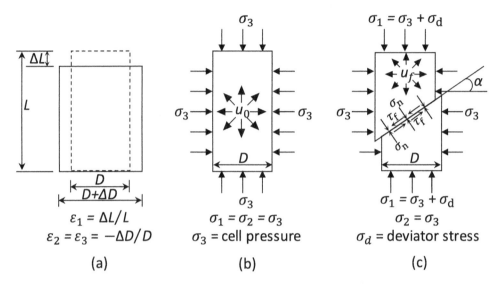

Figure 7.3 Strains and stresses in the triaxial test: (a) principal strains, (b) cell pressure only and (c) principal stresses at shear failure (modified after Whitlow, 1995).

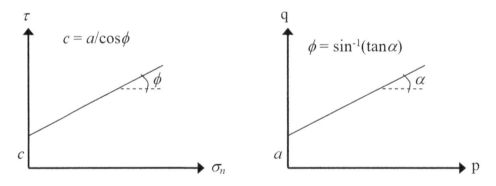

Figure 7.4 Shear strength enveloped plotted in the (a) τ-σ plane and (b) q-p plane.

For triaxial states, Eqs. 7.9 and 7.10 can be simplified due to the axial symmetry, or $\sigma_2 = \sigma_3$, so that previous equations can be simplified as follows:

$$p = \frac{\sigma_1 + 2\sigma_3}{3} \text{ or } p' = \frac{\sigma'_1 + 2\sigma'_3}{3} \tag{7.11}$$

$$q = \sigma_1 - \sigma_3 \text{ or } q = \sigma'_1 - \sigma'_3 \tag{7.12}$$

7.2.3 Pore pressure coefficients A and B

For undrained conditions, it is convenient to represent the variation of pore water pressure (Δu) associated with a change in the principal stresses ($\Delta\sigma_1$ and $\Delta\sigma_3$) using two pore pressure coefficients A and B as follows (Skempton, 1954):

$$\Delta u = B\left[\Delta\sigma_3 + A\left(\Delta\sigma_1 - \Delta\sigma_3\right)\right] \tag{7.13}$$

These coefficients can be measured experimentally in an undrained triaxial test, and the variation of $\Delta\sigma_1$ and $\Delta\sigma_3$ adopted typically depends on the problem under consideration. Typical values for coefficient A is shown in Table 7.1.

The increase in effective stress that results from an increase $\Delta\sigma_3$ for a test in partially saturated soil can be represented as follows:

$$\Delta\sigma' = \Delta\sigma_3 - \Delta u \tag{7.14}$$

Considering the compressibility of the soil skeleton (C_e) and compressibility of the fluid (C_f) in the voids (air and water), we can express the associated volume change of the specimen (ΔV_e) and volume change of the void space (ΔV_v) as:

$$\Delta V_e = -C_e V\left(\Delta\sigma_3 - \Delta u\right) \tag{7.15}$$

$$\Delta V_v = -C_f n V \Delta u \tag{7.16}$$

where V is the initial volume and n represents the porosity. For undrained conditions, these two changes in volume are identical ($\Delta V_e = \Delta V_v$) ; hence:

$$-C_e V\left(\Delta\sigma_3 - \Delta u\right) = -C_f n V \Delta u \tag{7.17}$$

Rearranging, we can obtain:

$$\frac{\Delta u}{\Delta\sigma_3} = B = \frac{1}{1+\dfrac{nC_f}{C_e}} \tag{7.18}$$

Table 7.1 Typical A values for different soils.

Soil type	A value
Very loose, fine sand	2–3
Sensitive clay	1.5–2.5
Normally consolidated clay	0.7–1.3
Lightly overconsolidated clay	0.3–0.7
Heavily overconsolidated clay	−0.5–0.0

For saturated soil, the compressibility of the water is very small compared to the compressibility of the soil skeleton, $C_f / C_e \approx 0$, and $B = 1$ when the degree of saturation is 1. In contrast for dry soil, the compressibility of the air is much larger than the compressibility of the soil skeleton, then $C_f / C_e \approx \infty$ and $B = 0$.

7.3 Undrained shear strength

The undrained shear strength of a soil is typically assessed using undisturbed specimens tested under unconsolidated and undrained conditions. Undisturbed specimens often have degree of saturation (S) less than 100% (unsaturated), which indicates pore water pressure is negative (i.e. suction) and total stresses are zero. With an increase in confining pressure, the pore water pressure in the specimen is increased, and the specimen becomes fully saturated. At this point, any increase in confining pressure will result in an equal increase in pore water pressure. Therefore, the Mohr circles corresponding to multiple specimens would have the same size, i.e. the principal stress difference is the same (Fig. 7.5). The results are plotted in total stresses, and the failure envelope is typically horizontal, in which the undrained shear strength (C_u) is given as follows:

$$C_u = \frac{q_f}{2} \tag{7.19}$$

where q_f is the principal stress difference at failure.

7.4 Stress paths

A stress path represents the locus of points representing the change of stress state of a soil, and it is typically represented in the $p'\text{–}q$ plane (analogous to $s'\text{–}t$ plane in 2D conditions). The total stress path (TSP) represents the conditions of the triaxial drained tests, whereas the effective stress path (ESP) represents the triaxial undrained tests with pore water pressure

$$q = \left(\sigma_1' - \sigma_3'\right) = (\sigma_1 - \sigma_3) \tag{7.20}$$

$$p' = \left(\sigma_1' + 2\sigma_3'\right)/3\left(\sigma_1 + 2\sigma_3\right)/3 - u = p - u \tag{7.21}$$

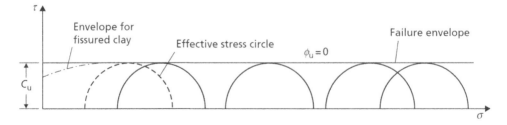

Figure 7.5 Unconsolidated undrained tests conducted for different confining pressures (after Knappett and Craig, 2012, reproduced with permission from CRC Press).

measurement, in which p' and q are adopted instead as follows and depicted in Figure 7.6.

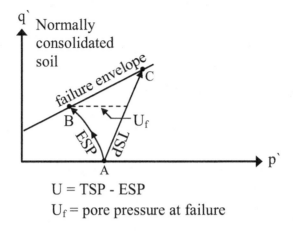

U = TSP - ESP

U_f = pore pressure at failure

Figure 7.6 Total and effective stress paths plotted in the q–p plane.

Worked-out example 7.1: A series of direct shear box tests was conducted in an undisturbed clay specimen under three different vertical stresses. The results obtained at failure are shown in Table 7.2.

Table 7.2 Test results.

σ_v (kPa)	200	300	400
τ_s (kPa)	95	130	162

a. Determine the apparent cohesion and the angle of friction (shearing resistance).
b. If another specimen is subjected to an undrained shear box test with vertical stress of 250 kPa, what would be the shear stress for which failure is expected?
c. If another specimen of this soil is subjected to an undrained triaxial test with confining pressure of 150 kPa, determine the total axial stress at which the failure would be expected.

Solution:
a. The problem may be solved either analytically or graphically.

Both methods are illustrated.
Graphic method (Fig. 7.7):
Measure the angle and cohesion intercept.

Cohesion intercept: 28 kPa
Angle of friction   18.5 degrees

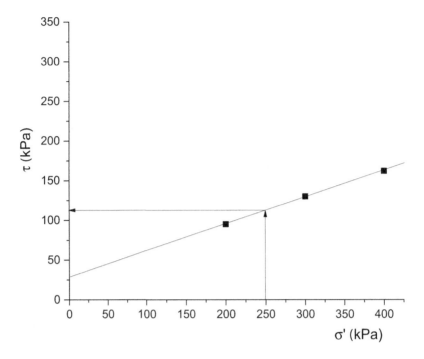

Figure 7.7 Test results of worked-out example 7.1.

Analytically,
Determine the slope of the failure line.

$$m = \frac{(162-95)}{(400-200)} = 0.335$$

$$\tan^{-1}(0.335) = 18.5°$$

Whereas the cohesion intercept can be found using equation of a line (e.g. $y = mx + b$):

$$\tau = c + m \times \sigma$$

$$95 = c + 0.335 \times 200$$

$$c = 28\,\text{kPa}$$

b. For a vertical stress of 250 kPa, then

$$\tau = 28 + 0.335 \times 250$$

$$\tau = 111.75\,\text{kPa}$$

c. From the relationship of the Mohr circle geometry, we have

$$\sigma_1 = \sigma_3 \left(\frac{1+\sin\phi}{1-\sin\phi} \right) + 2c\frac{\cos\phi}{1-\sin\phi}$$

If $\sigma_3 = 150 kPa$, $\phi = 18.5^{\circ}$ and $c = 28$ then

$$\sigma_1 = 150 \left(\frac{1+\sin 18.5}{1-\sin 18.5} \right) + 2 \times 28 \frac{\cos 18.5}{1-\sin 18.5} = 367.2\, kPa$$

The graphical solution is shown in Figure 7.8.

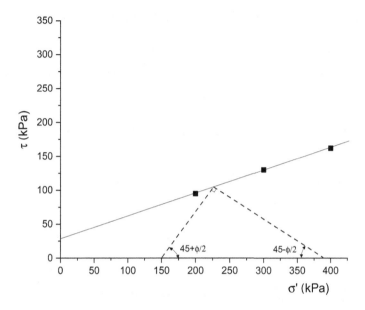

Figure 7.8 Graphical representation of worked-out example 7.1.

Worked-out example 7.2: A drained triaxial compression test carried out on three specimens of the same soil yielded the results displayed in Table 7.3. Draw the shear strength envelope and determine the shear strength parameters, assuming that the pore pressure remains constant during the axial loading stage.

Table 7.3 Drained triaxial compression test results.

Test no.	1	2	3
Cell pressure (kN/m²)	100	200	300
Ultimate deviator stress (kN/m²)	210	438	644

Solution: The ultimate principal stresses are obtained as follows:

Minor principal stress, σ_3 = cell pressure
Major principal stress, σ_1 = cell pressure + deviator stress
Since $U_f = 0$, $\sigma_1' = \sigma_1$ and $\sigma_3' = \sigma_3$

Test no.	1	2	3	
σ_3' = cell pressure (kPa)	100	200	300	Total stress = effective stress
σ_1' (axial stress) = $\Delta\sigma + \sigma_3'$	310	638	944	Because $\Delta U = 0$

Graphically, results are shown in Figure 7.9.

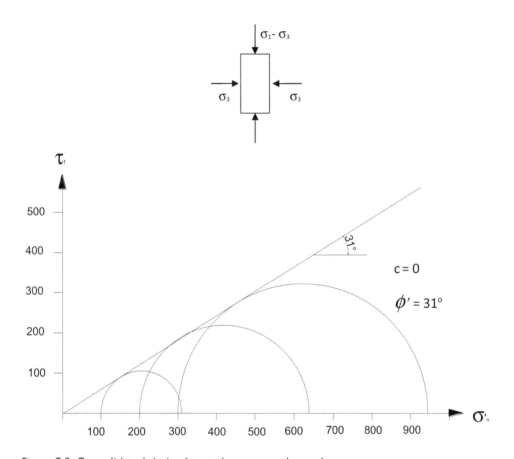

Figure 7.9 Consolidated drained test-shear strength envelope.

Worked-out example 7.3: The following results were obtained from consolidated undrained tests on specimens of saturated normally consolidated clay (Table 7.4). Determine

a. the effective stress parameters c' and ϕ' and
b. the apparent undrained strength parameters c and ϕ.

Table 7.4 Undrained triaxial compression test results.

Cell pressure (kN/m²)	75	150	300
Axial stress (kN/m²)	142	250	465
Excess pore pressure (kN/m²)	22	55	135

Solution:

a. Using the Mohr circle approach, three circles in terms of total stresses can be drawn. Calculations are shown in Table 7.5.

Table 7.5 Data obtained from tests results of worked-out example 7.3.

Confining pressure (σ_3) (kPa)	Axial stress (σ_1) (kPa)	Center of circle (x direction)	Center of circle (y direction)	Circle radius
75	142	108.5	0	33.5
150	250	200	0	50
300	465	382.5	0	82.5

The coordinates for the center of the Mohr circle can be obtained $(\sigma'_1 - \sigma'_3)/2 + \sigma'_3$ whereas the circle radius can be obtained considering $(\sigma'_1 - \sigma'_3)/2$

As this is an undrained test, three circles may be represented in terms of effective stresses, considering Table 7.6.

Table 7.6 Data obtained from tests results of worked-out example 7.3.

Effective confining pressure (σ'_3) (kPa)	Effective axial stress (σ'_1) (kPa)	Center of circle (x direction)	Center of circle (y direction)	Circle radius
53	120	86.5	0	33.5
95	195	145	0	50
165	330	247.5	0	82.5

Note that the radii of the circles are the same as those of total stresses because $\sigma_1' - \sigma_3' = \sigma_1 - \sigma_3$.

$$\sigma_1' = \sigma_1 - U_f$$
$$\sigma_3' = \sigma_3 - U_f$$

Graphical results are shown in Figure 7.10.

a. $c' = 5$ kPa, $\phi' = 18.3°$ (effective stress strength envelope)
b. $c = 15$ kPa, $\phi = 10.2°$ (total stress strength envelope)

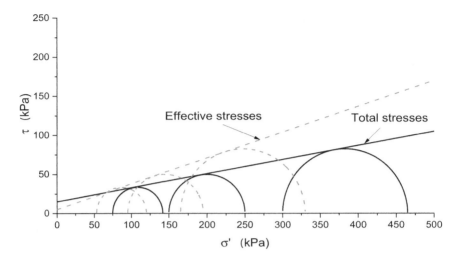

Figure 7.10 Mohr circles representation for worked-out example 7.3.

Worked-out example 7.4: In a consolidated-undrained triaxial test on a specimen of clay normally consolidated at a cell pressure of 130 kPa, the deviator stress and pore pressure at failure were recorded to be 250 kPa and 75 kPa, respectively. Determine the other corresponding shear strength parameters when:

a. $\phi_u = 0$, and
b. $c' = 0$

Solution: As per the information given,

$\sigma_3 = 130$ kPa, $U_f = 75$ kPa, then
$\sigma_3' = (130 - 75) = 55$ kPa
$\sigma_1' = 55 + 250 = 305$ kPa

a.   When $\phi_u = 0$, the envelope is parallel to the σ_n axis. The value of undrained cohesion can be found considering the radius of the Mohr circle, or $(\sigma_1' - \sigma_3')/2 = (305 - 55)/2 = 125$ kPa, thus $\tau_f = c_u = 125$ kPa.
b.   At failure $\sigma_3' = 55$ kPa and as $c' = 0$, the envelope passes through the origin. A simple trigonometric relationship can be used to determine the effective friction angle as follows:

$\sin \phi' = 125/180$ or
$\phi' = 44°$

Worked-out example 7.5: The preliminary site investigation for an excavation of tunnel reveals the presence of a porous sandstone with an unconfined compressive strength of

85 MPa. Triaxial compression tests conducted on the sandstone specimens collected indicate that its shear behavior can be modeled by a Mohr-Coulomb failure envelope, and a friction angle of 45° is obtained.

a. Determine the axial stress required to fail the sample in shear for a confining cell pressure of 5 MPa considering $\sigma_1 = \sigma_3 N\phi + 2c\sqrt{N\phi}$ and $N\phi = \dfrac{1+\sin 45°}{1-\sin 45°}$

b. If during the triaxial compression test the rubber jacket membrane is punctured and the pore pressure within the sample rises to the value of the cell pressure, what axial pressure is now required to cause failure?

Solution:

a. Given: $\sigma_c = 85$ MPa; $\phi = 45°$; $\sigma_3 = 5$ MPa

For $\phi = 45°$, $N_\phi = \dfrac{1+\sin 45°}{1-\sin 45°} = 5.83$

$\sqrt{N_\phi} = 2.414$

for unconfined conditions, $\sigma_3 = 0$ MPa, $\sigma_c = \sigma_{1f} = 85$ MPa, then

$\sigma_1 = \sigma_3 N_\phi + 2c\sqrt{N_\phi}$ can be simplified as follows:

$\sigma_{1f} = 2c\sqrt{N_\phi}$ or

$\sigma_c = 85$ MPa $= 2c\sqrt{N_\phi}$

Thus, $\sigma_3 = 5$ MPa; then using the general expression

$\sigma_{1f} = 5 \times 5.83 + 85 = 114.2$ MPa

$\sigma_{1f} = 114.2$ MPa

b. For $u_w = \sigma_3 = 5$ MPa $\rightarrow \sigma_3' = \sigma_3 - u_w = 0$ MPa

Effective stress $\rightarrow \sigma_1' = \sigma_3' N_\phi + 2c\sqrt{N_\phi}$

Assume that for rock $\phi' = 45°$ (no change) $\rightarrow \sigma_1' = 85$ MPa

$\sigma_1' = \sigma_1 - u_w$

$\sigma_1 = 85 + 5 = 90$ MPa

Worked-out example 7.6: A specimen was tested under consolidated drained conditions. At failure, the values shown in Table 7.7 were recorded. Determine the shear strength parameters of the soil tested.

Table 7.7 Undrained triaxial compression test results.

Cell pressure (kPa)	100	200	400
Axial stress (kPa)	310	578	1156

Solution:
Using the Mohr circle approach, three circles can be drawn (Table 7.8).

Table 7.8 Calculations for worked-out example 7.6.

Confining pressure (kPa)	Axial stress (kPa)	Center of circle (x direction)	Center of circle (y direction)	Circle radius
100	310	205	0	105
200	578	389	0	189
400	1156	778	0	378

The coordinates for the center of the Mohr circle can be obtained by $(\sigma_1' - \sigma_3')/2 + \sigma_3'$, whereas the circle radius can be obtained considering $(\sigma_1' - \sigma_3')/2$.

The circles can be drawn as shown in Figure 7.11. From the figure, it is clear that $c' = 0$, and friction angle can be easily calculated considering the sin trigonometric relationship between the center of the Mohr circle and its radius, for instance for the circle defined by 400 kPa confining pressure.

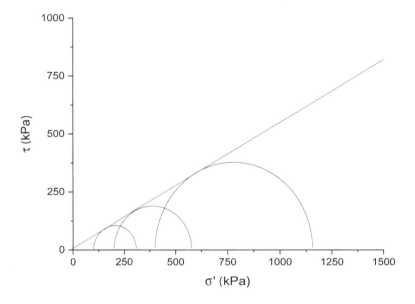

Figure 7.11 Mohr circle representation for the test data of worked-out example 7.6.

$$\sin\phi = \frac{378}{778}$$

$$\phi = 29°$$

Worked-out example 7.7: A reconstituted clay specimen (76 mm long and 38 mm diameter) is saturated and then allowed to consolidate under a cell pressure of 150 kPa. The drainage valve is then closed, and on increasing cell pressure to 300 kPa, the pore water pressure rises to 148 kPa. Determine the value of pore pressure parameter B and comment on whether the specimen has reached full saturation.

Solution:

$$\Delta u = B\left[\Delta\sigma_3 + A\left(\Delta\sigma_1 - \Delta\sigma_3\right)\right]$$

or

$$B = \frac{\Delta u}{\Delta\sigma_3 + A\left(\Delta\sigma_1 - \Delta\sigma_3\right)}$$

Based on the data given, there is an increase in pore water pressure of 148 for a variation in cell pressure of (300–150 = 150 kPa). As there is no additional axial stress applied to the specimen, the variation of σ_1 is the same as the variation of σ_3. Thus $\left(\Delta\sigma_1 - \Delta\sigma_3\right) = 0$.

Using the previous relationship, we have

$$B = \frac{148}{150 + A\left(0\right)} = 0.987$$

Typically values above $B = 0.95$ indicate that the specimen is fully saturated.

Worked-out example 7.8: The clay specimen (diameter 38 mm and height of 76 mm) described in worked-out example 7.7 is then subjected to triaxial compression, and the results obtained are given in Table 7.9.

Table 7.9 Undrained triaxial compression test results.

Axial displacement (mm)	Axial stress (kPa)	Pore water pressure (kPa)
0	300	148
2	532	242
4	591	244
6	609	228
8	610	215

a. Plot the deviator stress and excess pore water pressure against axial strain.
b. Determine the value of pore pressure parameter A for the different axial strain levels recorded.
c. Plot the total and effective stress paths for this test.

Solution:
a. Axial strain can be computed considering the initial height of the specimen (i.e. $h_0 = 76$ mm) as follows:

$$\varepsilon_a = \frac{\Delta h}{h_0}$$

Deviator stress can be calculated as follows:

$$\left(\sigma'_1 - \sigma'_3\right) = \left(\sigma_1 - \sigma_3\right)$$

And excess pore water pressures can be computed considering the current pore water pressure (u) and initial pore water pressure (u_0):

$$\Delta u = u - u_0$$

Then the data in Table 7.10 can be expressed in terms of axial strain, deviator stress and excess pore water pressure as follows, and graphical representation is shown in Figure 7.12.

Table 7.10 Calculations for worked-out example 7.8a.

Axial strain (%) $\varepsilon_a = \dfrac{\Delta h}{h_0}$	Deviator stress (kPa) $\left(\sigma'_1 - \sigma'_3\right) = \left(\sigma_1 - \sigma_3\right)$	Excess pore water pressure $\Delta u = u - u_0$ (kPa)
0.0	0	0
2.6	232	94
5.3	291	96
7.9	309	80
10.5	310	67

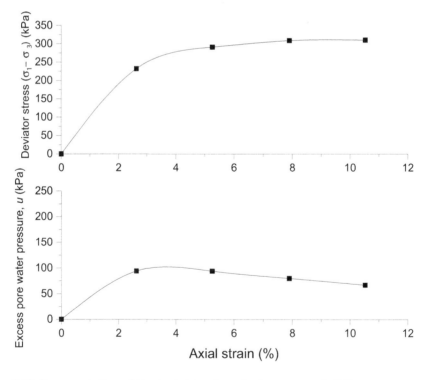

Figure 7.12 Representation of the test data of worked-out example 7.8.

b. The pore pressure parameter A can be computed for every level of axial strain considering the variation of the principal stresses (Table 7.11) as follows:

Table 7.11 Calculations for worked-out example 7.8b.

Axial strain (%)	Pore pressure parameter $A = \dfrac{\Delta u}{(\sigma_1 - \sigma_3)}$
0.0	0
2.6	0.4
5.3	0.33
7.9	0.26
10.5	0.22

$$\Delta u = B\left[\Delta\sigma_3 + A(\Delta\sigma_1 - \Delta\sigma_3)\right]$$

Since the all-round confining pressure produces a pore pressure $u = B \times \sigma_3$, and since we are considering the condition after shearing begins, the excess pore pressure is considered, then:

$$A = \frac{\dfrac{\Delta u}{B} - \Delta\sigma_3}{(\Delta\sigma_1 - \Delta\sigma_3)}$$

$$A = \frac{\Delta u}{(\sigma_1 - \sigma_3)}$$

c. The data given in Table 7.7 can be represented in terms of p and q as follows (Table 7.12, Fig. 7.13):

Table 7.12 Calculations for worked-out example 7.8c.

p (kPa)	p' (kPa)	q (kPa)
300.0	300.0	0
377.4	283.3	232
397.0	301.0	291
403.0	323.0	309
403.4	336.3	310

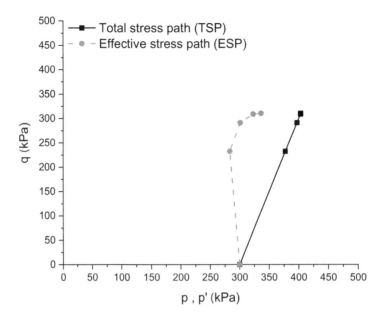

Figure 7.13 Representation of the stress paths of worked-out example 7.8c.

$$q = (\sigma_1 - \sigma_3) = (\sigma'_1 - \sigma'_3)$$

$$p' = p - \Delta u = \frac{(\sigma_1 + 2\sigma_3)}{3} - \Delta u$$

References

Knappett, J. & Craig, R.F. (2012) *Craig's Soil Mechanics*, 8th edition. CRC Press, ISBN:9780415561266, p. 584.

Skempton, A.W. (1954) The pore pressure coefficients A and B. *Geotechnique*, 4, 143–147.

Whitlow, R. (1995) *Basic Soil Mechanics*, 3rd edition. Longman Scientific & Technical, London, p. 559.

Chapter 8

Critical state soil behavior

This chapter focuses on the stress paths and critical state soil theory for different drainage conditions (Schofield and Wroth,1968; Wood,1990 and Knappett and Craig, 2012).

8.1 Stress paths and principal stresses

Where principal stresses act on horizontal and vertical planes (Fig. 8.1), $s' = \dfrac{\sigma'_v + \sigma'_h}{2}$ and $t = \dfrac{\sigma'_v - \sigma'_h}{2}$.

In Figure 8.2, AB is a line representing the locus of maximum shear stress (i.e. top of all the Mohr circles) until failure. Along AB, all points can be defined by coordinates (s, t) or $\left(\dfrac{\sigma_1 + \sigma_3}{2}, \dfrac{\sigma_1 - \sigma_3}{2} \right)$.

For triaxial stress states, it is also convenient to plot the p–q plane, where the y-axis is represented by deviatoric stress (as shown in Fig. 8.3) ($q = \sigma_1 - \sigma_3$ or $q' = \sigma'_1 - \sigma'_3$) and the x-axis is the mean stress: $p = \dfrac{\sigma_1 + 2\sigma_3}{3}$ or $p' = \dfrac{\sigma'_1 + 2\sigma'_3}{3}$

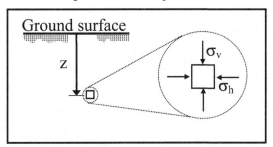

Figure 8.1 Illustration of principal stresses.

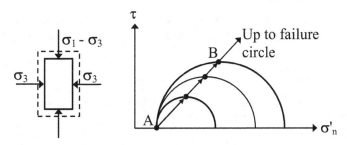

Figure 8.2 Illustration of Mohr-Coulomb plane and increase in principal stresses.

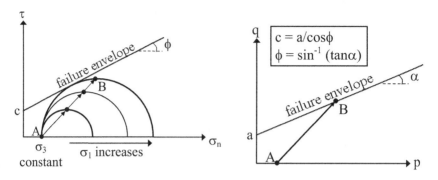

Figure 8.3 Stress path to failure.

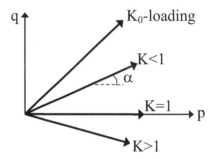

Figure 8.4 Illustration of different K_0 conditions.

8.1.1 Field loading conditions

8.1.1.1 Formation/deposition of soil layer

At K_0 condition, the lateral earth pressure coefficient can be computed as $K_0 = \dfrac{\sigma_h}{\sigma_v}$.

Initial condition $\sigma_1 = \sigma_3 = 0$, during deposition: $\sigma_h = K_0(\sigma_v)$, if $K_0 = 1$: isotropic compression (no shear stress) then $= \dfrac{q}{p} = \dfrac{3(1-K_0)}{1+2K_0}$. This is illustrated in Figure 8.4.

8.1.1.2 Compression and extension

In other field applications, the change in stress state typically involves compression and extension, as illustrated in Figure 8.5 and Figure 8.6 and in the q–p plane in Figure 8.7.

8.1.1.3 Other examples

There may be cases in which the initial state follows the isotopically consolidated stage ($\Delta\sigma_v = \Delta\sigma_h$ or $\Delta\sigma_1 = \Delta\sigma_3$), as shown in Figure 8.8.

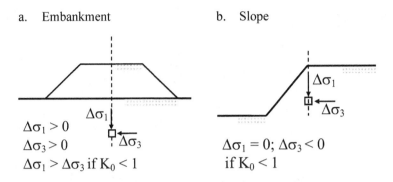

Figure 8.5 Illustration of principal stresses in (a) embankment and (b) slopes.

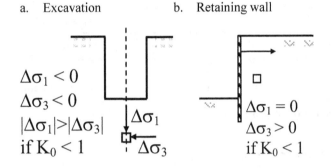

Figure 8.6 Illustration of principal stresses in (a) excavation and (b) retaining walls.

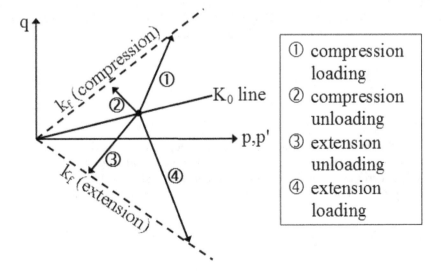

Figure 8.7 Illustration of main stress path combinations considering compression or extension and loading or unloading.

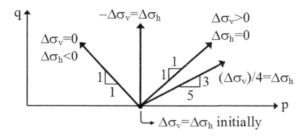

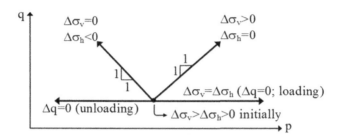

Figure 8.8 Illustration of main stress path combinations considering compression (loading) or extension (unloading) for specimens departing from the isotropic compression line (top) and from a certain (q, p) state.

8.2 Critical state theory

Critical state theory is an effective stress framework describing mechanical soil response under different loading conditions. It aims to describe the behavior of a saturated soil subjected to axisymmetric stress system using mathematical concepts. The basis of the theory is that if a soil is continuously distorted, it will attain a final failure state referred to as critical state, where its stress state and void ratio remains unchanged. This state is defined by two main equations, as follows:

$$q_f = Mp' \tag{8.1}$$

$$v = \Gamma - \lambda \ln p' \tag{8.2}$$

Where q_f is the deviatoric stress at failure (critical state), v is the specific volume ($v = 1 + e$), Γ, λ is the slope of the critical state line and M is shear stress ratio (q/p' at critical state).

The value of critical stress ratio (M) can be evaluated based on the final friction angle (ϕ_{cs}) measured by a conventional triaxial compression test, as shown in Figure 8.9.

$$M = \frac{6 \sin \phi_{cs}}{3 - \sin \phi_{cs}} \tag{8.3}$$

Eqs. 8.1 and 8.2 imply that critical state is achieved at a specific q/p' ratio and specific volume; thus a soil looser than critical will tend to contract before yielding if allowed to drain (pore pressure increases if undrained), whereas a soil denser than critical will tend to dilate if allowed to drain (pore pressure decreases if undrained). For a normally consolidated specimen, the initial consolidation stage can be illustrated in stress path OA (Fig. 8.10).

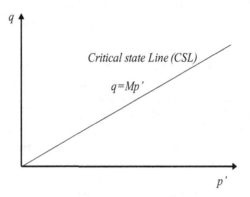

Figure 8.9 Illustration of the critical state line.

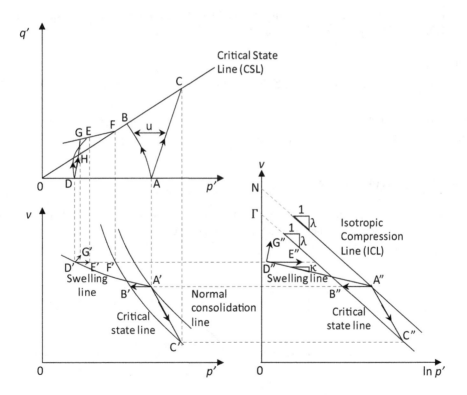

Figure 8.10 Illustration of main stress path combinations considering compression (loading) or extension (unloading)(modified after Sutton, 1993).

The points defined by different values of mean effective stress applied during a normally consolidated specimen define the isotropic compression line (ICL).

$$v_0 = N - \lambda \ln p' \tag{8.4}$$

Where v_0 is the specific volume at the initial consolidation pressure, N is the value of specific volume at $p' = 1$ kPa and λ is the slope of the isotropic compression line (typically the same as that of the critical state line).

For undrained and drained shearing, the stress path will continue to B and to C, respectively. If the specimen has been overconsolidated to A and then subsequently unloaded to D, then under undrained shearing, the stress path would continue to E and then F, whereas for drained shearing, it would continue to G, and as soil expands, it falls onto H on the critical state line (CSL).

Similar observations of these paths and associated volumetric changes can be made if the same stress paths are plotted in terms of $v - p'$ and $v - \ln p'$ space (Fig. 8.10).

In addition, soil states above and below the CSL are often referred to as being on the "wet" side and "dry" side, respectively.

Worked-out example 8.1: A specimen was collected and then subjected to isotropic consolidation (i.e. all-around cell pressure). The cell pressure level adopted was 40 kPa. The specimen was then subjected to additional load of 100 kPa.

a.   Determine the deviatoric stress and mean effective stress during the consolidation stage.
b.   Draw the stress path for this specimen.

Solution:

a.   Using $s = \dfrac{\sigma_1 + \sigma_3}{2}$ and $t = \dfrac{\sigma_1 - \sigma_3}{2}$,

at the initial stage with all-around cell pressure of 40 kPa ($\sigma_1 = \sigma_3 = 40$ kPa), then

$$s = \frac{40 + 40}{2} = 40 = 40 \text{ and } t = \frac{40 - 40}{2} = 0$$

Point A $(s, t) =$ A (40, 0) (kPa).

b.   With the loading, there is an increase σ_1 to 100 kPa, while $\sigma_3 = 40$ is kept constant (i.e. cell pressure, then the soil state moves to point B $(s, t) =$ B (70,30) (Fig. 8.11).

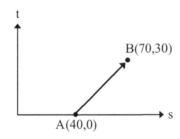

Figure 8.11 Illustration of stress path for worked-out problem 8.1.

Worked-out example 8.2: A specimen was collected in a site-investigation campaign, and the in-situ stresses (vertical and horizontal) measured were $\sigma_v = 80$ kPa, $\sigma_h = 60$ kPa. A foundation structure was constructed; subsequently, the in-situ stresses measured $\sigma_v = 120$ kPa, $\sigma_h = 70$ kPa. Draw the stress path for this condition. If there a stress relief and vertical and horizontal stresses are reduced to 100 kPa and 65 kPa, respectively, plot the change in stress path.

Solution:

While in most triaxial tests, stress paths start from the horizontal axis, in the field $\sigma_1 \neq \sigma_3$, hence, $t = 0$ condition does not exist. For this problem, the in-situ condition is $\sigma_v = 80$ kPa, $\sigma_h = 60$ kPa.

Then, using $s = \dfrac{\sigma_v + \sigma_h}{2}$ and $t = \dfrac{\sigma_v - \sigma_h}{2}$, the coordinates of point A representing the initial condition are as follows:

$A(s, t) = A(70, 10)$

Due to foundation loading: $\sigma_v = 120$ kPa, $\sigma_h = 70$ kPa

→ $B(s, t) = B(95, 25)$

With stress relief (unloading): $\sigma_v = 100$ kPa and $\sigma_h = 65$ kPa.
then $C(s, t) = C(82.5, 17.5)$. The stress path changes (Fig. 8.12).

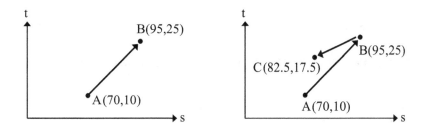

Figure 8.12 Illustration of stress path for worked-out problem 8.2.

Worked-out example 8.3: A new road embankment is proposed. To evaluate the shear strength of the soil, a sample is consolidated in a triaxial cell, adopting a confining stress of 200 kPa to mimic field conditions. Determine the mean effective stress, the effective major principal stress and the void ratio at failure when shearing is carried out under drained conditions simulating long-term analysis. The critical state soil parameters for this soil determined in an additional set of tests are $\Gamma = 2.2$, $\lambda = 0.1$, $M = 0.77$ and $N = 2.5$.

Solution:

For drained conditions, the specific volume at the end of consolidation can be found by

$v_0 = N - \lambda \ln p'$ or

$$v_0 = 2.5 - 0.1 \ln 200 = 1.97$$

and

$$v_0 = 1 + e_0$$
$$e_0 = 0.97$$

Since the confining stress is constant, the stress path will have a slope of 3:1 until it reaches the critical state line ($q = 0.77p'$). Then,

$$q_f = Mp'$$
$$q_f = 0.77 p'_f$$
$$p'_f = 269.06 \text{ kPa}$$

and $q_f = 0.77 p'_f = 207.18 \text{ kPa}$

If $q = \sigma'_1 - \sigma'_3$ and $p' = \dfrac{\sigma'_1 + 2\sigma'_3}{3}$, then

$$p'_f = \frac{(\sigma'_{1f} + 2\sigma'_{3f})}{3} = 269.06 \text{ kPa}$$
$$q'_f = (\sigma'_{1f} - \sigma'_{3f}) = 207.18 \text{ kPa}$$

Then, solving for σ'_{1f}, we can obtain

$$\sigma'_{1f} = 407.17 \text{ kPa}$$

To determine the void ratio at failure, the relationship that describes the CSL can be used as follows:

$$v = \Gamma - \lambda \ln p'$$
$$v = 2.2 - 0.1 \ln 270.3$$
$$v = 1.64$$
$$e = 0.64$$

Worked-out example 8.4: For the site conditions outlined in worked-out problem 8.3 and considering short term stability (undrained conditions), determine the deviatoric stress, mean effective stress and the void ratio at critical state (i.e. failure).

Solution:
In undrained shearing, no volume change takes place during shearing, then the initial void ratio is the same as the final void ratio at failure, as follows:

$$v_0 = N - \lambda \ln p' \text{ or}$$
$$v_0 = 2.5 - 0.1 \ln 200 = 1.97$$

and

$$v_0 = 1 + e_0$$
$$e_0 = 0.97$$

To determine the mean effective stress at failure, the CSL relationship may be used as follows Atkinson and Bransby (1978):

$$v_f = \Gamma - \lambda \ln p'_f$$
$$1.97 = 2.2 - 0.1 \ln p'_f$$
$$p'_f = 9.97 \, \text{kPa}$$

And $q_f = Mp'_f$

$$q_f = 0.77 p'_f = 0.77 \times 9.97 = 7.6 \, \text{kPa}$$

Worked-out example 8.5: An excavation is planned in a sand deposit. The critical friction angle of the sand was determined via triaxial compression tests to be 32°. Determine the value of the critical shear ratio M and the final failure strength of the sand if the mean effective stress at failure is 100 kPa.

Solution:
The value of the critical shear ratio M can be computed based on the critical friction angle as follows:

$$M = \frac{6 \sin \phi_{cs}}{3 - \sin \phi_{cs}}$$

$$M = \frac{6 \sin 32}{3 - \sin 32} = 1.29$$

As $q_f = Mp'_f$ then

$$q_f = 1.29 p'_f$$
$$q_f = 1.29 \times 100 = 129 \, \text{kPa}$$

References

Atkinson, J.H and Bransby, P. L. (1978) *The Mechanics of Soils: An Introduction to Critical State Soil Mechanics*, McGraw Hill, London.

Knappett, J and Craig, R.F. (2012) *Craig's Soil Mechanics*, 8th Edition, CRC Press, ISBN: 9780415561266.

Schofield, A and Wroth, P. (1968) *Critical State Soil Mechanics*, McGraw Hill, London.

Sutton, B.H.C. (1993) *Solving Problems in Soil Mechanics*. Longman Scientific & Technical, Harlow, UK, p. 267, ISBN-10: 0582089719.

Wood, D. M (1990) *Soil Behavior and Critical State Soil Mechanics*, Cambridge University Press, ISBN 0-521-33249-4.

Chapter 9

Bearing capacity of foundations

Foundations are part of civil infrastructure that facilitate transfer of both static and dynamic loads from the superstructure safely to the underlying ground. Shallow foundations are considered when geological material at the ground surface has ample strength to withstand this applied load. Deep foundations are usually required when the soil below the structure has a relatively poor load-bearing capacity; thus, the loads must then be carried to deeper soil layers using piles (driven or bored), granular columns or caissons. In general, the bearing capacity can be defined as the largest intensity of applied pressure by a structural member to the soil, which supports it without causing excessive settlement or shear failure. In view of design, the ultimate and allowable bearing capacities play an important role.

Ultimate bearing capacity (q_{ult}): the maximum pressure a foundation soil can withstand without causing shear failure in soil.

Allowable bearing pressure (q_a): the maximum allowable net loading pressure at which the soil neither fails in shear nor exhibit excessive settlement.

9.1 Bearing capacity equations

A foundation can be considered shallow when the depth of the foundation is less than its width (D < B).

The theory proposed by Terzaghi (1943) is commonly used for determining the bearing capacity of a foundation. It was originally developed for a strip footing considering the roughness of the foundation and its self-weight below the base, hence:

$$q_{ult} = cN_c + \gamma DN_q + 0.5B\gamma N_\gamma \tag{9.1}$$

Where N_c, N_q and N_γ are non-dimensionless bearing capacity factors that depend only on the angle of internal friction (ϕ). γ is the unit-weight of the soil, c is the cohesion of the soil, D is the depth of the foundation and B is the width of the foundation.

Terzaghi (1943) bearing-capacity factors are given by:

$$N_c = \left(N_q - 1\right)\cot\phi \tag{9.2}$$

$$N_q = \frac{a^2}{2\cos^2\left(45° + \dfrac{\phi}{2}\right)} \quad \text{where } a = e^{\left(0.75\pi - \frac{\phi}{2}\right)\tan\phi} \tag{9.3}$$

$$N_\gamma = \frac{\tan\phi}{2}\left(\frac{K_{p\gamma}}{\cos^2\phi} - 1\right)$$ (9.4)

Where $K_{p\gamma}$ is the passive earth pressure coefficient and angles are given in degrees.

Figure 9.1 shows the variation of Terzaghi's bearing capacity factors with the friction angle.

The general form of bearing-capacity equation proposed by Meyerhof (1963) is:

$$q_{ult} = cN_c i_c d_c s_c + \gamma D N_q i_q d_q s_q + 0.5B\gamma N_\gamma i_\gamma d_\gamma s_\gamma$$ (9.5)

Where i, d and s are load inclination, depth and shape factors, respectively. The bearing capacity factors for N_c, N_q and N_γ for Meyerhof (1963) and Hansen (1970) and Vesic (1973) are shown in Figure 9.2, and their algebraic forms are given by:

$$N_c = \left(N_q - 1\right)\cot\phi$$ (9.6)

$$N_q = \tan^2\left[45 + \frac{\phi}{2}\right]e^{\pi\tan\phi}$$ (9.7)

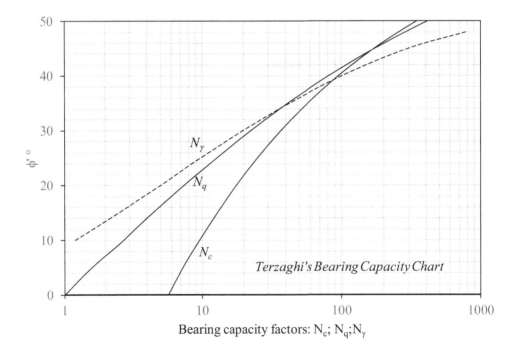

Figure 9.1 Terzaghi's bearing-capacity factors for shallow foundations (modified after Terzaghi and Peck, 1967).

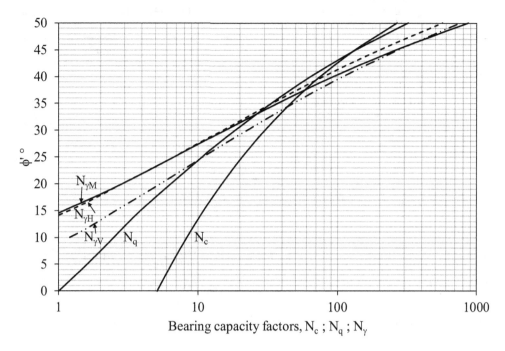

Figure 9.2 Bearing capacity factors (modified after Knappett and Craig, 2012).

The various expressions for N_γ can be obtained from:

$$\text{Meyerhof}\ (1963): N_{\gamma M} = \left(N_q - 1\right)\tan\ (1.4\phi) \tag{9.8}$$

$$\text{Hansen}\ (1970): N_{\gamma H} = 1.5\ (N_q - 1)\tan\phi \tag{9.9}$$

$$\text{Vesic}\ (1973): N_{\gamma V} = 2\ (N_q + 1)\tan\phi \tag{9.10}$$

The shape, depth and load inclination factors for Terzaghi, Hansen and Meyerhof are tabulated in Table 9.1.

9.1.1 Effect of water table on bearing capacity

If the water table is at the base of the footing,

$$q_{ult} = cN_c + \gamma DN_q + 0.5B\gamma'N_\gamma \tag{9.11}$$

γ' = effective unit weight of the soil within the failure zone
   If the water table is at the ground surface, then:

$$q_{ult} = cN_c + \gamma'DN_q + 0.5B\gamma'N_\gamma \tag{9.12}$$

Table 9.1 Shape, depth and inclination factors for Terzaghi (1943), Hansen (1970), Meyerhof (1963) and Vesic (1973, 1975).

	Shape factor	Depth factor	Inclination factor
Terzaghi (1943)	$S_c = 1.0$ (strip); 1.3 (round); 1.3 (square) $S_\gamma = 1.0$(strip); 0.6 (round); 0.8 (square)	–	–
Hansen (1970)	$S_c = 1 + \dfrac{N_q}{N_c}\dfrac{B'}{L'}$ $B' = B - 2e_x \quad L' = L - 2e_y$ $S_c = 0.2\dfrac{B'}{L'}$ for $\phi = 0°$ $B' = B - 2e_x \quad L' = L - 2e_y$ $S_q = 1 + \dfrac{B'}{L'}\sin\phi$ $B' = B - 2e_x \quad L' = L - 2e_y$ $S_\gamma = 1 - 0.4\dfrac{B'}{L'} \geq 0.6$ $B' = B - 2e_x \quad L' = L - 2e_y$	$d_c = 1 + 0.4\dfrac{D}{B}$ for $\dfrac{D}{B} \leq 1$ $d_c = 0.4\dfrac{D}{B}$ for $\phi = 0°$ $d_c = 1 + 0.4\,\tan^{-1}\left(\dfrac{D}{B}\right)$ for $\dfrac{D}{B} \geq 1$ $d_q = 1 + 2\tan\phi\left(1 - \sin\phi\right)^2 {D}/{B}$ for $\dfrac{D}{B} \leq 1$ $d_q = 1 + 2\tan\phi\left(1 - \sin\phi\right)^2 {D}/{B}$ for $\dfrac{D}{B} \leq 1$ $d_\gamma = 1$ for all ϕ	$i_c = 1 - \dfrac{H}{2cB'L'}$ $i_q = 1 - 1.5\dfrac{H}{V}$ $i_\gamma = i_q^2$
Meyerhof (1963)	$S_c = 1 + 0.2 k_p\dfrac{B}{L}$ $k_p = \tan^2\left(45° + \dfrac{\phi}{2}\right)$ $S_q = S_\gamma = 1 + 0.1 k_p\dfrac{B}{L}\phi \geq 10°$ $S_q = S_\gamma = 1 \quad \phi = 0$	$d_c = 1 + 0.2\sqrt{k_p}\dfrac{D}{B}$ Any ϕ $k_p = \tan^2\left(45° + \dfrac{\phi}{2}\right)$ $d_q = d_\gamma = 1 + 0.1\sqrt{k_p}\dfrac{D}{B}\phi \geq 10°$ $d_q = d_\gamma = 1 \quad \phi = 0$	$i_c = i_q = \left(1 - \dfrac{\theta}{90}\right)^2$ Any ϕ $i_\gamma = \left(1 - \dfrac{\theta}{\phi°}\right)^2 \phi > 0$ $i_\gamma = 0\,\theta > 0\,\phi = 0$
Vesic (1973, 1975)	$S_c = 1 + \dfrac{N_q}{N_c}\dfrac{B}{L} = 1$ for stip $S_q = 1 + \dfrac{B}{L}\tan\phi$ $S_\gamma = 1 - 0.4\dfrac{B}{L} \geq 0.6$	Same as Hansen	$i_c = 1 - \dfrac{mH_i}{A_f C_a N_c}$ for $\phi = 0$ $i_c = i_q - \dfrac{1 - i_q}{N_q - 1}$ for $\phi > 0$ $i_q = \left[1 - \dfrac{H_i}{V + A_f C_a \cot\phi}\right]^m$ $i_\gamma = \left[1 - \dfrac{H_i}{V + A_f c_a \cot\phi}\right]^{m+1}$ $m = m_B = \dfrac{(2 + B/L)}{(1 + B/L)}$ $m = m_L = \dfrac{(2 + L/B)}{(1 + L/B)}$

If the water table is well below the foundation level (e.g. D > 1.5B), the bulk unit weight can be used for determining q_{ult}.

9.1.2 Net ultimate bearing capacity (q_{net}) and factor of safety (F)

$$q_{net} = q_{ult} - \gamma D \tag{9.13}$$

Factor of safety (*F*) with respect to shear failure is defined as:

$$F = \frac{q_{ult} - \gamma D}{q - \gamma D} \tag{9.14}$$

where *V* and *H* are the vertical and horizontal components of the resultant load, respectively.
θ = Angle of resultant measured from vertical without a sign.

9.2 Standard penetration tests

Standard in-situ penetration test is widely used to determine an approximate value for the allowable bearing pressure. The standard penetration number (*N*) is the blow count required to drive the standard split barrel sampler over a depth of 300 mm into the ground. The details of the penetration test can be found in BS 1377 (part 9). Terzaghi and Peck (1967) developed an empirical chart corroborating the allowable bearing pressure, the standard penetration number (*N*) and the foundation width for a limiting maximum settlement to 25 mm (Fig. 9.3).

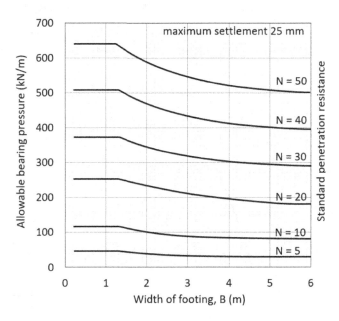

Figure 9.3 Design chart for allowable bearing pressure (modified after Terzaghi and Peck, 1967).

9.2.1 Overburden correction

In addition to density index, standard penetration resistance depends on the effective stress at the depth of the measurement, and it can be approximated by the effective overburden pressure (Fig. 9.4).

9.2.2 Dilatancy correction

If N value is greater than 15, for fine sand and silty sand below the water table, this should be corrected for the resistance due to negative excess pore pressure during in-situ testing.

$$N' = 15 + 0.5(N - 15)$$
(9.15)

9.2.3 Correction for water table

If the water table is at the ground level, estimated q_a should be reduced by 50%.

If the water table depth (D_w) is $\geq$ to the width (B) below the foundation level, no reduction is required.

If the water table depth (D_w) is between ground level and width (B) below the foundation depth, then the value of q_a should be multiplied by the correction factor, C_w:

$$C_w = 0.5 + 0.5 \frac{D_w}{D + B}$$
(9.16)

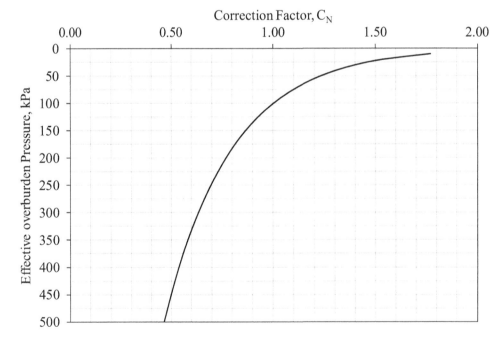

Figure 9.4 Correction for overburden pressure (modified after Peck, Hanson and Thornburn, 1974).

Where D_w is the depth of water table below the ground surface.

Meyerhof (1965) suggested the following equations for calculating the allowable bearing pressure considering the measured value of N.

$$q_a = \frac{S_L N}{1.9} \quad \text{for } B < 1.25 \text{ m} \tag{9.17a}$$

$$q_a = \frac{S_L N}{2.84} \left(\frac{B + 0.33}{B} \right)^2 \quad \text{for } B > 1.25 \text{ m} \tag{9.17b}$$

Where S_L is the permitted settlement limit (mm); N = average N value between $z = D$ and $z = D + B$ and B = width of the foundation.

9.3 Pile foundations

There are two basic types of pile foundation: (1) end bearing piles and (2) friction piles.

End bearing piles transmit load through weak layers of deposit to a firm stratum which is capable of carrying the load.

Friction piles mainly derive the support from the frictional resistance developed between the pile and soil.

The ultimate bearing resistance (Q_u) of a single pile is the sum of ultimate base resistance (Q_b) and ultimate skin friction (Q_s).

$$Q_u = Q_b + Q_s \tag{9.18}$$

$$Q_u = A_b q_b + A_s q_s \tag{9.19}$$

Where q_p is the ultimate bearing capacity of soil at the base of the pile; A_p is the base area of the pile; q_s is the ultimate shearing resistance between soil and pile; and A_s is the perimeter area of the shaft.

9.3.1 q$_p$ and q$_s$ for sand

$$q_b = \sigma_0' N_q \quad \text{(neglect term } N_\gamma \text{ due to the small } B \text{ compared to } D) \tag{9.20}$$

$$q_s = k_s \sigma_0' \tan \delta \tag{9.21}$$

k_s is the average coefficient of earth pressure along the embedded length; σ_0' is the average effective overburden pressure along the embedded length; δ is the angle of friction between pile and sand.

9.3.2 q$_p$ and q$_s$ for clay

$$q_b = c_u N_c \tag{9.22}$$

N_c is the bearing capacity factor, and equal to 9 is widely accepted for design.
c_u is the undrained shear strength of undisturbed clay at the base of the pile.

$$q_s = \alpha \bar{c}_u \qquad (9.23)$$

α is the adhesion factor depending on the type of clay, pile material and method of installation. The values range from 0.3 to 1.0. $\bar{c}_u$ is the average undisturbed undrained shear strength of clay adjoining the pile.

9.3.3 q_p and q_s from in-situ method (Meyerhof, 1976)

Driven piles in sand and gravel:

$$q_b = 40N \frac{D_b}{B} \leq 400N \, (\text{kPa}) \qquad (9.24)$$

Driven piles in non-plastic silts:

$$q_b = 40N \frac{D_b}{B} \leq 300N \, (\text{kPa}) \qquad (9.25)$$

Bored piles in granular soil:

$$q_b = 14N \frac{D_b}{B} \, (\text{kPa}) \qquad (9.26)$$

Where N is the average value of uncorrected N value in the vicinity of pile base; B is the width of the pile; D_b is the length of the pile embedment in the soil.

Driven piles (large diameter) $q_s = 2\bar{N}$ (9.27)

Driven piles (average diameter) $q_s = \bar{N}$ (9.28)

Bored piles $q_s = 0.67\bar{N}$ (9.29)

Where $\bar{N}$ is the average of the uncorrected N value over the embedded length of the pile.

9.4 Pile groups

A pile foundation is generally constructed as a group of piles and connected to the top to a pilecap. The structural load is applied to the pile cap. The ultimate load-carrying capacity of a pile group cannot be always the summation of the individual carrying capacities of individual piles due to a phenomenon called group action. Failure of the group pile may be due to failure of the individual pile or to failure of the overall block supporting the group.

Efficiency: The ratio of the average load in a pile group to the ultimate single-pile load is referred to as the efficiency (E) of the group.

$$E = \frac{\text{ultimate group load}}{\text{N} \times \text{ultimate individual pile load}} \qquad (9.30)$$

for driven pile in sand $E = 1$ and for bored piles in sand $E < 0.6$.

9.4.1 Block failure

The block failure is an important characteristic of pile groups in clayey soils.

$$Q_u = A_b q_b + A_s c_s \qquad (9.31)$$

Where A_b is the base area of the group; A_s is the perimeter area of the group; c_s is the average value of shearing resistance/unit area along the sides of the pile; q_b is the ultimate bearing resistance at the base of the pile.

9.4.2 Settlement of pile group

Due to the effect of group action, the total settlements (both immediate and consolidation) of the group will be greater than for a single pile.

End bearing piles: The total foundation load is assumed to act at the base of the piles on a virtual foundation of the same size as the plan of the pile foundation.

Friction piles: It is assumed that the structural load is effectively transferred to a depth of 2/3 of the penetration depth. It is further assumed that there is a distribution of the structural load of 1H:4V. The settlement of this equivalent foundation can be considered the settlement of the group.

Worked-out example 9.1: A square foundation is 2 m wide and is placed at a depth of 2.8 m in a soil having $c' = 10$ kPa; $\phi' = 20°$ and $\gamma = 19$ kN/m³. Determine the ultimate bearing capacity using the

i.   Terzaghi,
ii. Meyerhof and
iii. Hansen approaches

Solution:
i.   Terzaghi approach:

$q_{ult} = cN_c + \gamma D N_q + 0.5 B \gamma N_\gamma$

From Table 9.1: $s_c = 1.3$; $s_q = 1.0$; $s_\gamma = 0.8$

From Figure 9.1: $N_c = 17.7$; $N_q = 7.4$; $N_\gamma = 5.0$

$q_{ult} = 1.3 \times 10 \times 17.7 + 19 \times 2.8 \times 7.4 + 0.4 \times 2 \times 19 \times 5 = 700 \, \text{kpa}$

ii. Meyerhof (1963) approach:

$q_{ult} = cN_c d_c s_c + \gamma D N_q d_q s_q + 0.5 B \gamma N_\gamma d_\gamma s_\gamma$

From Figure 9.2: $N_c = 14.8$; $N_q = 6.4$; $N_\gamma = 2.9$

Shape factors:

From Table 9.1: $s_c = 1.41$; $s_q = 1.2$; $s_\gamma = 1.2$

Depth factors:

From Table 9.1: $d_c = 1.40$; $d_q = 1.2$; $d_y = 1.2$

$q_{ult} = (10 \times 14.8 \times 1.40 \times 1.41) + (19 \times 2.8 \times 6.4 \times 1.2 \times 1.2) + (0.5 \times 2 \times 19 \times 2.9 \times 1.2 \times 1.2)$

$q = 861.79 \, \text{kPa}$

iii. Hansen approach:

$q_{ult} = cN_c d_c s_c + \gamma DN_q d_q s_q + 0.5B\gamma N_y d_y s_y$

From Figure 9.2: $N_c = 14.8$; $N_q = 6.4$; $N_y = 2.9$

Shape factors:

From Table 9.1: $s_c = 1.43$; $s_q = 1.34$; $s_y = 0.6$

Depth factors:

From Table 9.1: $d_c = 1.38$; $d_q = 1.3$; $d_y = 1.0$

$q_{ult} = (10 \times 14.8 \times 1.38 \times 1.43) + (19 \times 2.8 \times 6.4 \times 1.3 \times 1.34) + (0.5 \times 2 \times 19 \times 2.9 \times 1 \times 0.6)$

$q_{ult} = 918.2 \, \text{kPa}$

Worked-out example 9.2: A square footing (2 m × 2 m) is to be founded on a sandy soil ($\phi = 30$) at a depth of 1.5 m. The unit weights of sand above and below the water table are 18 kN/m³ and 19 kN/m³, respectively. Determine the ultimate bearing capacity of the footing when the water table is (a) at the ground surface and (b) at the base of the foundation.

Solution:

From Figure 9.2: $\phi = 30°$: $N_c = 30.1$; $N_q = 18.4$; $N_y = 15.7$ (Meyerhof's factor)

a. At ground level:

$q_{ult} = cN_c d_c s_c + \gamma' DN_q d_q s_q + 0.5B\gamma' N_y d_y s_y$

$q_{ult} = 0 + (19 - 9.81) \times 1.5 \times 18.4 \times 1.13 \times 1.3 + 0.5 \times 2 \times 15.7 \times (19 - 9.81) \times 1.13 \times 1.3$

$q_{ult} = 585 \, \text{kPa}$

b. At the base of the foundation:

$q_{ult} = cN_c d_c s_c + \gamma D\gamma DN_q d_q s_q + 0.5B\gamma' N_y d_y s_y$

$q_{ult} = 0 + 18 \times 1.5 \times 18.4 \times 1.13 \times 1.3 + 0.5 \times 2 \times 15.7 \times (19 - 9.81) \times 1.13 \times 1.3$

$q_{ult} = 942 \, \text{kPa}$

Worked-out example 9.3: A square footing (2.5 m × 2.5 m) is 2 m below the ground surface. The water table is far below the base level. The vertical (V) and horizontal (H) components of the base reaction are 300 and 100 kN/m, respectively. The eccentricity of the base reaction is 0.2 m. The unit weight of the soil is 17.5 kN/m³, and the effective shear strength parameters are $c' = 0$ and $\phi = 32$. Determine the ultimate bearing capacity of the foundation.

Solution:

From Figure 9.2 : $\phi = 32°$: $N_c = 35.5$; $N_q = 23.2$; $N_\gamma = 20.8$ (Hansen's factor)

$B' = B - 2e = 2.5 - 2 \times 0.2 = 2.1$ m

Shape factors: Assuming the eccentricity was in the direction of B:

From Table 9.1 : $s_q = 1.45$; $s_\gamma = 0.66$

Depth factors:

From Table 9.11 : $d_q = 1.22$; $d_\gamma = 1.0$

Inclination factors:

From Table 9.1 : $i_q = 0.5$; $i_\gamma = 0.25$

$q_{ult} = \gamma D N_q d_q i_q s_q + 0.5 B' \gamma N_\gamma d_\gamma i_\gamma s_\gamma$

$q_{ult} = 17.5 \times 2 \times 23.2 \times 1.22 \times 0.5 \times 1.45 + 0.5 \times 2.1 \times 17.5 \times 20.8 \times 1 \times 0.25 \times 0.66$

$q_{ult} = 781.3$ kPa

Worked-out example 9.4: A strip footing founded at a depth of 1 m is designed to support a load of 600 kN/m to dense sand layer having the following properties: $c' = 0$, $\phi = 36°$ and $\gamma_{sat} = 20.5$ kN/m³.

Assuming that the water table may rise to the ground surface and adopting a factor of safety of 3.0, determine the breadth of footing (neglect shape and depth factors).

Solution:

$q_{ult} = \gamma' D N_q + 0.5 B \gamma' N_\gamma$

From Figure 9.2 : $\phi = 36°$: $N_c = 50.5$; $N_q = 37.7$; $N_\gamma = 40.0$ (Hansen's factor)

$q_{ult} = (20.5 - 9.81) \times 1 \times 37.7 + 0.5 \times B \times (20.5 - 9.81) \times 40 = 403 + 214B$

$= 3F = \dfrac{q_{ult} - \gamma D}{q - \gamma D} = \dfrac{403 + 214B - 10.69}{\dfrac{600}{B} - 10.69} = 3$

$214B^2 + 424.38B - 1800 = 0$

$B = 2.07$ m

Worked-out example 9.5: A square footing of 2.5 m × 2.5 m is placed 2 m deep within a sand deposit. The water table is 2 m below the ground surface. The measured standard penetration resistance N is recorded below. Determine the allowable bearing capacity for a limiting settlement of 25 mm based on (a) Terzaghi and Peck (1943) and (b) Meyerhof (1965) approaches (assume $\gamma = 19$ kN/m³ and $\gamma_{sat} = 21$ kN/m³).

Depth (m)	1.5	2.5	3.5	4.5	5.5	6.5	7.5	8.5
N value	8	9	14	15	23	28	32	32

Solution:

a. Terzaghi and Peck (1967) suggested considering N values from the base of the foundation to a depth equal to B below the square foundation (see table below).

Depth (m)	Nvalue	σ'_v (kPa)	C_N (from Fig. 9.4)	N (corr.) = (N × C_N)
1.5	8	-	-	-
2.5	9	43.6	1.28	12
3.5	14	54.8	1.2	17
4.5	15	66	1.14	17
5.5	23	77.17	1.09	25
6.5	28	88.36	1.04	29
7.5	32	-	-	-
8.5	32	-	-	-

The average value of N interpolated between depths 2.5 m and 6.5 m is 20. The provisional value of allowable bearing capacity is 210 kPa (from Fig. 9.3), which should then be multiplied by C_w to correct for the effect of the water table. Thus:

$$C_w = 0.5 + 0.5\frac{D_w}{D+B}$$

$$= 0.5 + 0.5\frac{2}{2+2.5} = 0.72$$

Therefore, the allowable bearing pressure is given by $q_a = 0.72 \times 210 = 151$ kPa.

b. Meyerhof (1965) approach:

The average value of measured N between $z = D$ and $D + B$ is 15. Hence:

$$q_a = \frac{25 \times 15}{2.84}\left(\frac{2.5+0.33}{2.5}\right)^2 = 169 \text{ kPa}$$

This example shows that the Terzaghi and Peck method yields more conservative values of q_a compared to the Meyerhof approach.

Worked-out example 9.6: A single pile of diameter 500 mm and length 10 m is to be installed into a deep clay deposit. The undrained shear strength of 150 kPa was determined at a depth of 10 m, and an average undrained shear strength of 110 kPa was recorded for 0 to 10 m. Assuming $\alpha = 0.6$, determine the ultimate bearing capacity of the pile.

Solution:

$$Q_u = A_b q_b + A_s q_s$$

$$q_b = c_u N_c = 150 \times 9 = 1350 \text{ kPa}$$

$$A_b = \frac{\pi}{4} \times 0.5^2 = 0.196 \text{ m}^2$$

$$q_s = \alpha \bar{c}_u = 0.6 \times 110 = 66 \text{ kPa}$$

$$A_s = \pi \times 0.5 \times 10 = 15.7 \text{ m}^2$$

$$Q_u = 0.196 \times 1350 + 15.7 \times 66 = 1301 \text{ kN}$$

Worked-out example 9.7: A 6-m-thick layer of sand overlies a deep deposit of dense gravel. A series of standard penetration tests on sand has shown an average N value of 21. A precast pile of square section 0.25 m × 0.25 m is driven through sand. Adopting a factor of safety of 3.0, determine the allowable bearing capacity of the pile.

Solution:

$$Q_u = A_b q_b + A_s q_s$$

$$q_b = 40N \frac{D_b}{B} \leq 400 N$$

$$q_b = 400N = 400 \times 21 = 8400 \text{ kPa}$$

$$A_b q_b = (0.25)^2 \times 8400 = 525 \text{ kN}$$

$$q_s = \bar{N} = 21$$

$$A_s q_s = 6 \times 0.25 \times 4 \times 21 = 126 \text{ kN}$$

$$Q_u = 525 + 126 = 651 \text{ kN}$$

$$Q_{allowable} = \frac{Q_u}{FoS} = \frac{651}{3} = 217 \text{ kN}$$

Worked-out example 9.8: A single pile with 450 mm diameter has been driven 15 m into a soft clay, which has an undrained cohesion of 20 kN/m² and a unit weight of 18 kN/m³. The groundwater table is found to be at the surface of the clay. Determine the ultimate pile load capacity (assume $\alpha = 0.95$ and $N_c = 9$).

Solution:

$$Q_u = A_b q_b + A_s q_s$$

$$A_b = \frac{\pi}{4} 0.45^2 = 0.16 \text{ m}^2$$

$$A_s = \pi \times 0.45 \times 15 = 21.2 \text{ m}^2$$

$$q_b = c_u N_c = 20 \times 9 = 180 \text{ kPa}$$

$$q_s = \alpha \bar{c}_u = 0.95 \times 20 = 19\,\text{kPa}$$

$$Q_u = 0.16 \times 180 + 21.2 \times 19 = 432\ \text{kN}$$

References

British Standard 1377 (1990) *Methods of Test for Soils for Civil Engineering Purposes*. British Standards Institution, London.

Hansen, J. (1970) *A Revised and Extended Formula for Bearing Capacity*. Danish Geotechnical Institute, Bulletin 28, Copenhagen.

Knappett, J. & Craig, R.F. (2012) *Craig's Soil Mechanics*, 8th edition. CRC Press, ISBN:9780415561266, p. 584.

Meyerhof, G.G. (1963) Some recent research on the bearing capacity of foundations. *Canadian Geotechnical Journal*, 1(1), 16–26.

Meyerhof, G.G. (1965) Shallow foundations. *Journal of Soil Mechanics and Foundation Division, ASCE*, 92(2), 21–31.

Meyerhof, G.G. (1976) Bearing capacity and settlement of pile foundations. *Proceedings ASCE*, 102(GT3), 195–228.

Peck, R.B., Hanson, W.E. & Thornburn, T.H. (1974) *Foundation Engineering*. John Wiley & Sons, New York.

Terzaghi, K. (1943) *Theoretical Soil Mechanics*. John Wiley & Sons, New York.

Terzaghi, K. & Peck, R.B. (1967) *Soil Mechanics in Engineering Practice*, 2nd edition, John Wiley & Sons, New York.

Vesic, A.S. (1973) Analysis of ultimate loads of shallow foundations. *Journal ASCE*, 99(SM1), 45–73.

Vesic, A.S. (1975) *Principles of Pile Foundation Design*, Soil Mechanics Series No. 38, School of Engineering, Duke University, Durham, NC.

Chapter 10

Mass movement and slope stability analysis

Instability of road and rail cuttings, sliding of rock on jointed surfaces such as bedding planes and fault lines, soil and rock movement in open-cut mines, coastal slope instability caused by bottom erosion (loss of passive support) and opening of rock fractures, large-scale flow of saturated soil (debris flows) and other forms of soil and rock mass movements define various geometries of ground instability. These movements can be classified as finite and infinite slope movements that can be further subdivided with respect to circular, noncircular, translational and complex geometry of the moving mass. This chapter introduces different aspects related to the stability of such slopes and the analytical methods for determining the factor of safety (FOS). If the computed FOS is greater than unity, the slope is considered safe, and if FOS is less than unity, the slope is considered to be unsafe.

10.1 Rotational or circular slips –total stress analysis (undrained)

The analysis in terms of total stresses (no pore water pressures considered) or immediate undrained conditions aim to simulate site conditions where fully saturated clay is subjected to short-term instability (e.g. immediately after construction). In total stress analysis, the moment equilibrium across a circular arc segment is conveniently considered (Fig. 10.1), and here the potential instability is driven mainly by the self-weight of the soil mass (W), and the resistance to failure is predominantly a function of the shear force mobilized on the circular slip surface (S_{mob}), where C_u is the undrained cohesion acting along the arc length (AB).

The factor of safety may be computed considering the moment equilibrium about the point of rotation O, as follows:

$$\text{Factor of safety (F)} = (\Sigma \text{ mobilized resisting moments})/(\Sigma \text{ disturbing moments}) \quad (10.1)$$

where

$$\Sigma \text{ mobilized resisting moments} = S_{mob} \times R = C_u \times AB \times R \quad (10.2)$$
$$\Sigma \text{ Disturbing moments} = W \bar{x} \quad (10.3)$$

Then,

$$F = C_u \times AB \times R = W \bar{x} \text{ or } F = \frac{C_u R^2 \theta}{W \bar{x}} \quad (10.4)$$

Note that at failure, $F = 1$.

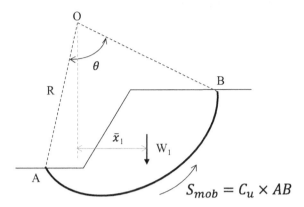

Figure 10.1 Forces acting on a circular slip slope.

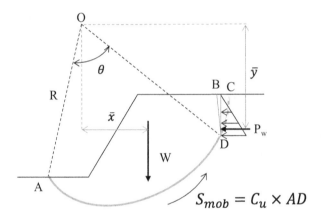

Figure 10.2 Forces acting on a circular slip slope.

10.1.1 Effect of the tension crack

If there is a tension crack (B-D-C) filled with water located on the top of the slope, then the area ABC is deducted from total weight of wedge, and the water level in the crack contributes to a disturbing moment (Fig. 10.2).

$$\Sigma \text{ mobilized resisting moments} = S_{mob} \times R = C_u \times AB \times R \tag{10.5}$$

$$\Sigma \text{ Disturbing moments} = W\bar{x} + P_w\bar{y} \tag{10.6}$$

The factor of safety, F, is now given by:

$$F = \frac{C_u \times AB \times R}{W\bar{x} + P_w\bar{y}} \tag{10.7}$$

In general, a water-filled tension crack reduces the FOS. In practice, the cracks formed due to soil movement need to be sealed or covered to protect water infiltration and thereby cause slope instability.

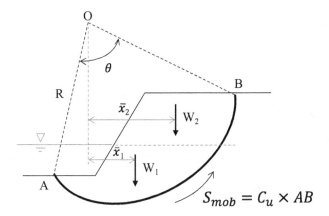

Figure 10.3 Forces acting on a partially submerged circular slip slope.

10.1.2 Effect of submergence

If only a section of the slope is submerged, then the water level can be extended, and the submerged unit weight ($\gamma_{sub} = \gamma_s - \gamma_w$) is used instead to compute the weight force for the area of the slope below the water table (Figure 10.3).

$$\Sigma \text{ Mobilized resisting moments} = S_{mob} \times R = C_u \times AB \times R \qquad (10.8)$$

$$\Sigma \text{ Disturbing moments} = W_1 \times \overline{x_1} + W_2 \times \overline{x_2} \qquad (10.9)$$

10.1.3 Taylor's stability chart

The factor of safety can also be related to the slope (geometrical) properties, i.e. H (height of the slope) and β (inclination angle of the slope) and the distinct characteristics related to the slip surface geometry (R and θ). Taylor (1937) published design stability charts for the cases in which C_u is uniform with depth, and a rigid boundary is imposed at a given depth (Fig. 10.4). Gibson and Morgenstern (1962) published charts that incorporate a linear variation of C_u with depth. In either of these methods, no external loads or the effect of tension cracks or pore water pressure are considered. The stability factor (N) can be defined by Eq. 10.10, where the numerator represents the shearing resistance while the denominator represents the driving or disturbing force related to slope geometric factors, where F is the factor of safety.

$$N = \frac{c_u}{F \gamma H} \qquad (10.10)$$

10.2 Rotational or circular slips – effective stress analysis (drained)

In this analysis, the slope is divided into different slices of width b and height h_i (measured at the center of slice), and the base of the slice is assumed to be straight, having l_i length (Fig. 10.5). The net forces on vertical boundaries ($E_n - E_{n+1}$) and ($X_n - X_{n+1}$) shown in Figure 10.5 exist for any particular slice, and the problem is statically indeterminate.

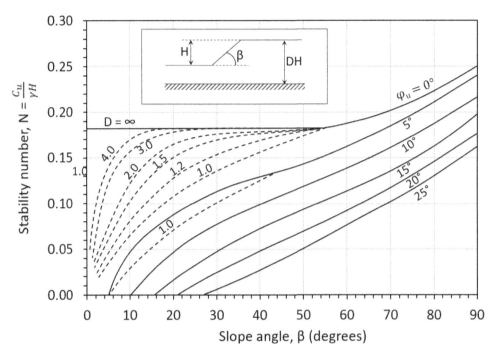

Figure 10.4 Taylor's stability chart (modified after Whitlow, 1993).

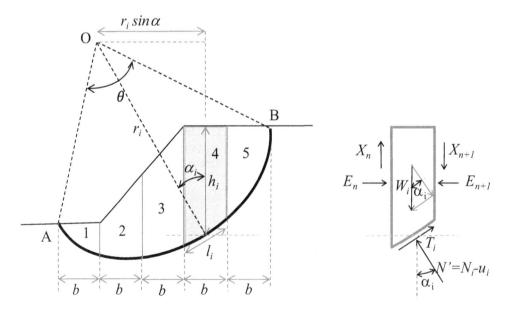

Figure 10.5 Forces acting on a circular slip slope.

10.2.1 Method of slices – Swedish (conventional) method (Fellenius, 1936)

This method assumes that the result of net force on vertical boundaries $(E_n - E_{n+1})$ and $(X_n - X_{n+1})$ is zero. The factor of safety is based on the moment equilibrium about point O considering the disturbing and resisting forces normal to the base of ith slice, as follows:

$$\text{Factor of safety (F)} = (\Sigma \text{mobilized resisting moments})/(\Sigma \text{ disturbing moments}) \quad (10.11)$$

or

$$F = \frac{\sum T_i r_i}{\sum W_i\, r_i \sin \alpha_i} \, or \, F = \frac{\sum T_i}{\sum W_i \sin \alpha_i} \tag{10.12}$$

where $T_i = \tau\, l_i$ and τ represents the shear strength along the length l_i, which for drained conditions is given by effective stresses (i.e. $\tau = c' + \sigma_i' \tan\phi'$), then

$$F = \frac{\sum \left(c' + \sigma_i' \tan\phi'\right) l_i}{\sum W_i \sin \alpha_i} \tag{10.13}$$

or

$$F = \frac{c'b\sec\alpha_i + W_i\left(\cos\alpha_i - r_{u,i}\sec\alpha_i\right)\tan\phi'}{\sum W_i \sin\alpha_i} \tag{10.14}$$

Where $l_i = b \sec\alpha$, $W = \gamma\, b\, h_i$ and $r_{u,i}$ is the ratio of the pore pressure to the overburden pressure at the base of the slice.

For many problems, this method is sufficiently accurate, but for deep-seated circles it has been shown to underestimate the factor of safety with errors up to 20% owing to the assumptions. Hence caution must be exercised when using it in practice.

10.3 Method of slices – Bishop's simplified method

Considering the balance of forces in the vertical direction (Fig. 10.5), we have:

$$W_i + \left(X_n - X_{n+1}\right) = \left(N_i - u_i\, l_i\right)\cos\alpha_i + \left(N_i - u_i\, l_i\right)\frac{\tan\phi'}{F}\sin\alpha_i$$
$$+ u_i\, l_i \cos\alpha_i - \frac{c'}{F}\sin\alpha_i \tag{10.15}$$

Solving for $(N - u_i\, l_i)$, then,

$$\left(N - u_i l_i\right) = \frac{W_i + \left(X_n - X_{n+1}\right) - u_i l_i \cos\alpha_i + \dfrac{c'}{F}\sin\alpha_i}{\cos\alpha_i + \dfrac{\tan\phi'}{F}\sin\alpha_i} \tag{10.16}$$

As $\sigma'_i = N - u_i l_i$ then replacing in Eq. 10.11,

$$F = \frac{\sum\left(c' + \left[\dfrac{W_i + (X_n - X_{n+1}) - u_i l_i \cos\alpha_i + \dfrac{c'}{F}\sin\alpha_i}{\cos\alpha_i + \dfrac{\tan\phi'}{F}\sin\alpha_i}\right]\tan\phi'\right) l_i}{\sum W_i \sin\alpha_i} \tag{10.17}$$

In Eq. 10.17, there is no assumption made in relation to $(E_n - E_{n+1})$, as this term does not influence the forces vertically. Hence, Bishop (1955) proposed a method to consider nonzero term $(X_n - X_{n+1})$; however, this refinement only provides marginal effect on the factor of safety where thin slices are considered. Rearranging Eq. 10.17 and considering $(X_n - X_{n+1}) = 0$ for a typically thin slice and $l_i = b\sec\alpha$, $W = \gamma b h_i$ and $r_{u,i} = u_i \times \sec\alpha/W$, we obtain:

$$F = \frac{\sum\left(\left[c'b + W_i(1 - r_{u,i})\tan\phi'\right]\dfrac{\sec\alpha_i}{1 + \dfrac{\tan\phi'}{F}\tan\alpha_i}\right)}{\sum W_i \sin\alpha_i} \tag{10.18}$$

As the factor of safety appears on both sides of Eq. 10.18, it must be determined via iteration, i.e. a trial value of F is chosen and then successively improved. The factor of safety determined by this method tends to be underestimated, but the error is unlikely to exceed about 7% and in most cases is less than 2%. Moreover, the method assumes that the values of F for both cohesion and frictional resistance are the same.

10.4 Bishop and Morgenstern charts

Based on Eq. 10.19, Bishop and Morgenstern (1960) proposed dimensionless stability coefficients (m and n) for homogenous slopes, and it was shown that the factor of safety varies linearly with r_u as follows:

$$F = m - n r_u \tag{10.19}$$

Where m and n can be computed based on the slope angle and soil properties (Fig. 10.6).

10.5 Finite planar slope analysis – sliding block analogy

Figure 10.7 shows an example profile. Consider 1 m length along AB plane where $C' = c' \times$ AB (per unit length), e.g. bedding planes are often filled with clay and silt.

$$C'_{mob} = \frac{c'l}{F} \tag{10.20}$$

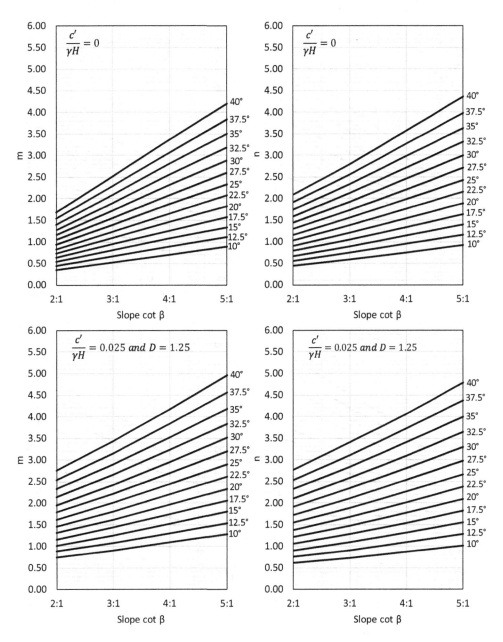

Figure 10.6 Bishop and Morgenstern's dimensionless stability coefficients (modified after Bishop and Morgenstern, 1960).

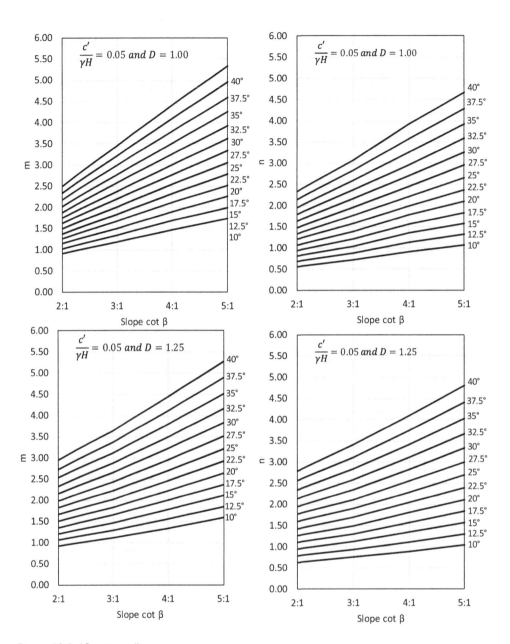

Figure 10.6 (Continued)

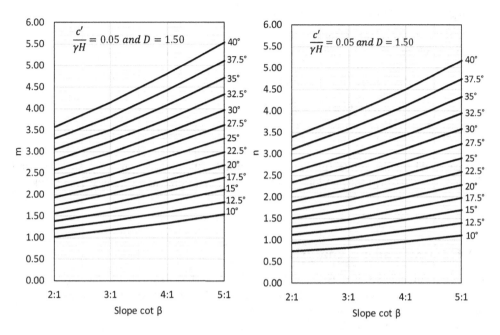

Figure 10.6 (Continued)

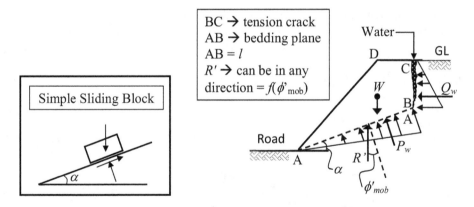

Figure 10.7 Forces acting on finite plane slope.

Resolving forces in the direction parallel to the AB plane gives:

$$W \sin \alpha + Q_w \cos \alpha = C'_{mob} + R' \sin \phi'_{mob} \tag{10.21}$$

Resolving of forces in the direction perpendicular to the AB plane gives:

$$W \cos \alpha - Q_w \sin \alpha = P_w + R' \cos \phi'_{mob} \tag{10.22}$$

R' may be eliminated considering $\tan\phi'_{mob} = \dfrac{\tan\phi'}{F}$.

Then the factor of safety for this finite planar slope can be computed based on the limit equilibrium between the resisting and disturbing forces, where:

F = (Σ resisting forces)/(Σ disturbing forces), hence:

$$F = \frac{c'l + (W\cos\alpha - Q_w\sin\alpha - P_w)\tan\phi'}{W\sin\alpha + Q_w\cos\alpha} \tag{10.23}$$

For cases where the joints (plane AB) are filled with granular material such as sand (i.e. $u = 0$, $c' = 0$), then the previous equation can be simplified as a simple sliding block:

$$F = \tan\phi'/\tan\alpha \tag{10.24}$$

10.6 Infinite slope analysis

In this type of analysis, it is assumed that the failure surface is parallel to the surface of the slope and that the sliding mass thickness is considerably smaller compared to the length of the slope. Thus the slope can be considered to have an infinite length (Fig. 10.8).

The shear strength mobilized at the base of the slope:

$$\tau = c' + \sigma'_n \tan\phi' \text{ where } \sigma'_n = N'/l \tag{10.25}$$

Resolving forces in the direction parallel to the slip surface gives:

$$W\sin\beta = S = \tau \cdot l/F \tag{10.26}$$

Considering $W = b\,z\,\gamma = l\,z\,\gamma\cos\beta$, then:

$$\therefore F = \frac{c' + (N'/l)\tan\phi'}{\gamma z\sin\beta\cos\beta} \tag{10.27}$$

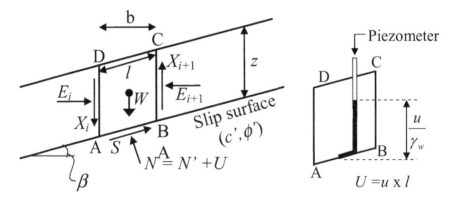

Figure 10.8 Forces acting on infinite plane slope.

Resolving forces in the direction perpendicular to the slip surface gives:

$$N = N' + U = W \cos \beta \tag{10.28}$$
$$\therefore N' = l\left(\gamma z \cos^2 \beta - u\right)$$

Combining Eq. 10.27 and Eq. 10.28 yields:

$$F = \frac{c' + \left(\gamma z \cos^2 \beta - u\right)\tan \phi'}{\gamma z \sin \beta \cos \beta} \tag{10.29}$$

In terms of total stresses, $\tau = c_u$; $\phi_u = 0$

$$\therefore F = \frac{S}{W \sin \beta} = \frac{c_u \cdot l}{\gamma z b \sin \beta} \quad \rightarrow \quad F = \frac{c_u}{\gamma z b \sin \beta \cos \beta} \tag{10.30}$$

10.6.1 Selected special cases

i. $c' = 0$ (e.g. sliding on existing shear surface)

$$F = \left(1 - \frac{u}{\gamma z \cos^2 \beta}\right)\frac{\tan \phi'_u}{\tan \beta} \tag{10.31}$$

ii. $c' = 0$, $u = 0$ (sliding block)

$$F = \tan \phi'/\tan \beta \tag{10.32}$$

iii. Horizontal flow lines (Fig. 10.9): In this case, replace u by:

$$u = z\,\gamma_w \text{ or } r_u = \gamma_w\gamma \tag{10.33}$$

where $r_u = u/\gamma z$

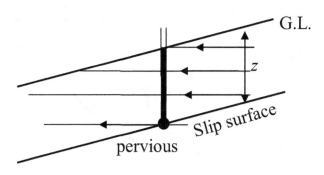

Figure 10.9 Horizontal flow lines on an infinite plane slope.

iv.   Flow lines parallel to ground surface (Fig. 10.10): AB is an equipotential line drawn perpendicular to flow lines.

$$u = \gamma_w \cdot mz \cos^2 \beta$$

$$\therefore F = \left(1 - \frac{m\gamma_w}{\gamma}\right) \frac{\tan \phi'}{\tan \beta} \tag{10.34}$$

If $m = 1$, groundwater level (GWL) and ground level (GL) coincide; thus $u = \gamma_w z \cos^2\beta$.

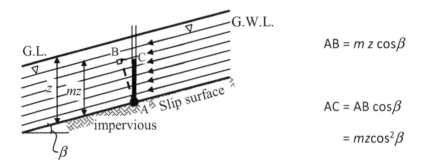

Figure 10.10 Flow lines parallel to the infinite plane.

Worked-out example 10.1: The bank of a canal has the profile shown in Figure 10.11. The material is a homogenous clay with a unit weight of 20 kN/m³, undrained cohesion of 30 kPa, and it is fully undrained; hence, assume $\phi_u = 0$. For the trial slip circle shown, the area ABCD is 150 m² and the centroid is at W_1. A tension crack (CD) of 2.5 m deep was also observed. Calculate the FOS for two different scenarios shown in Figure 10.11 and Table 10.1.

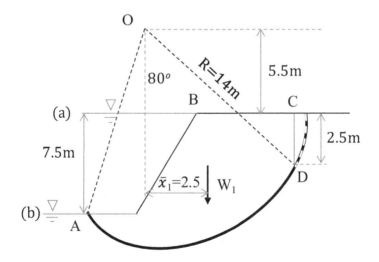

Figure 10.11 Circular slip slope.

Table 10.1 Scenarios

Scenario	Water level in the canal	Depth of the water level inside the tension crack (m)
(a)	Full – 7.5 m	Full – 2.5 m
(b)	Empty	Full – 2.5 m

Solution:

a. When the canal is full, the soil can be considered to be submerged, and the submerged unit weight is then used to determine the factor of safety. For this condition, the water pressure in the tension crack is balanced by the water pressure of the water in the canal. Considering 1 m run of the bank and taking moments about Point O:

Factor of safety (FOS) = (Σ mobilized resisting moments)/(Σ disturbing moments)

where
Σ Disturbing moments = $W\bar{x}$
Σ Disturbing moments = $(20-9.81) \times 150 \times 2.5 = 3821.25$ kN/m
Σ mobilized resisting moments = $C_u \times AD \times R$

$$\Sigma \text{ mobilised resisting moments} = 30 \times 14 \times \frac{\pi \times 80}{180} \times 14 = 8210.03 \text{ kN.m}$$

Then

$$\text{Factor of safety } (F) = \frac{8210.03}{3821.25} = 2.15$$

b. When the canal is empty, the saturated unit weight is used instead. In addition, the water pressure in the tension crack adds to the disturbing moment component; thus:

$$\Sigma \text{ Disturbing moments} = W\bar{x} + P_w\bar{y}$$
$$\Sigma \text{ Disturbing moments} = 20 \times 150 \times 2.5 + 9.81 \times \frac{2.5^2}{2} \times \left(5.5 + \frac{2}{3} \times 2.5\right)$$
$$= 7500 + 219.70 = 7719.7 \text{ kN.m}$$

The mobilized resisting moments remain the same as follows:

$$\Sigma \text{ mobilized resisting moments} = C_u \times AD \times R$$
$$\Sigma \text{ mobilised resisting moments} = 30 \times 14 \times \frac{\pi \times 80}{180} \times 14 = 8210.03 \text{ kN.m}$$

Then

$$\text{Factor of safety } (F) = \frac{8210.03}{7719.7} = 1.06$$

Worked-out example 10.2: For the previous exercise regarding a canal bank, consider the canal is one third full (Fig. 10.12). Compute the factor of safety for this condition, considering

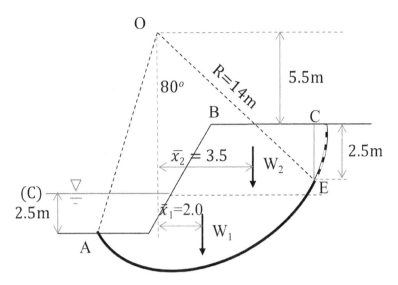

Figure 10.12 Circular slip slope.

that the area below the water level is 90 m² and the area above the water level is 60 m². The tension crack is assumed to be filled with water.

Solution:
When the canal is partially full to one-third, there are two components of the weight forces above (saturated unit weight) and below (submerged unit weight) the water level. Considering 1 m run of the bank and taking moments about Point O, we obtain:

Factor of safety (F) = (Σ mobilized resisting moments)/(Σ disturbing moments)

where
Σ Disturbing moments = $W_1 \times \bar{x}_1 + W_2 \times \bar{x}_2$
Σ Disturbing moments = (20–9.81) × 90 × 2.0 + 20 × 60 × 3.5 = 6034.2 kN.m
Σ mobilized resisting moments = C_u × AE × R
Σ mobilised resisting moments = $30 \times 14 \times \dfrac{\pi \times 80}{180} \times 14 = 8210.03$ kN · m

Then,

$$\text{Factor of safety}\,(F) = \frac{8210.03}{6034.2} = 1.36$$

Worked-out example 10.3: A vertical road cutting is proposed on a site that consists of a saturated clay with a saturated unit weight of 17.2 kN/m³ and an average undrained shear strength c_u of 33 kN/m².
Determine the maximum height for which the cutting may be temporarily unsupported.

Solution:

In this problem, Taylor's stability charts can be used to determine the maximum unsupported height. For the condition of failure, the factor of safety is 1.0.

From the chart (Fig. 10.13) for $\beta = 90°$ (vertical cut) and $\phi = 0$ (saturated clay), then $N = 0.25$.

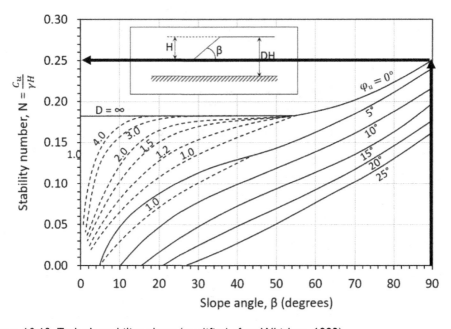

Figure 10.13 Taylor's stability chart (modified after Whitlow, 1993).

Then the maximum unsupported height can be computed considering Taylor's stability factor as follows:

$$N = \frac{c_u}{F\gamma H} \text{ rearranging for H gives}$$

$$H = \frac{C_u}{F\gamma N}$$

$$H = \frac{33}{1 \times 17.2 \times 0.25} = 7.67 \text{ m}$$

Worked-out example 10.4: A new cutting for a road having a height of 5.5 m is proposed on a site having 11.1-m-thick saturated clay, followed by intact shale as bedrock. The saturated unit weight of the clay is 19.5 kN/m³, and the undrained shear strength is 28 kPa.

a.   Determine the maximum side slope angle against failure for a factor of safety of 1.5.
b.   Determine the factor of safety if both during and subsequent to excavation, the cutting is kept flooded with fresh water (i.e. to original ground level).

Solution: The problem can be illustrated as shown in Figure 10.14.

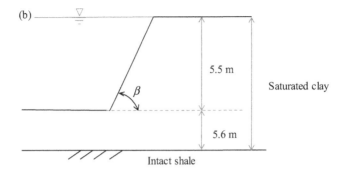

Figure 10.14 Worked-out example 10.4 diagram illustration.

a.   Using Taylor's stability charts,

$$H = 5.5 \text{ and } D = \frac{11.1}{5.5} = 2$$

$$N = \frac{c_u}{F\gamma H} = \frac{28}{1.5 \times 19.5 \times 5.5} = 0.174$$

Considering $D = 2$ curve, for $N = 0.174$, a maximum angle of the slope of 24° is obtained.

b.   For fully flooded conditions, use $\gamma_{sub} = \gamma_{sat} - \gamma_w$. Then

$$N = \frac{c_u}{F\gamma H} = \frac{28}{F \times (19.5 - 9.81) \times 5.5} = \frac{0.525}{F}$$

Considering $\beta = 24°$ and $D = 2.0$, N is the same as the in (a) or $N = 0.174$.
Then F = 0.525 / 0.174 = 3.02.
A greater factor safety is obtained for fully flooded conditions.

Worked-out example 10.5: A cutting 20 m deep at a 3:1 slope is proposed on a site composed of an over consolidated clay. Laboratory drained triaxial tests results revealed that this soil has $c' = 14$ kPa, $F = m - n\,r_u = 1.8 - 1.65(0.15) = 1.55$ $\phi' = 25°$, and it has a bulk unit weight of 19.2 kN/m³ and a pore pressure ratio, $r_u = 0.15$. Determine the factor of safety of the cutting using the Bishop and Morgenstern (1960) stability coefficient charts.

Solution: The problem can be illustrated as shown in Figure 10.15.

The angle of the slope = 3:1 or $\beta = \tan^{-1}\left(\frac{1}{3}\right) = 18.4°$.

$$\frac{c'}{\gamma H} = \frac{14}{19.2 \times 20} = 0.036$$

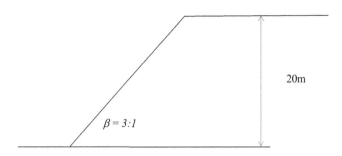

Figure 10.15 Worked-out example 10.5 diagram illustration.

From the Bishop and Morgenstern (1960) method, as there is no chart for $\dfrac{c'}{\gamma H} = 0.036$, the factor of safety must be interpolated between two available values.

So consider the nearest values of $\dfrac{c'}{\gamma H} = 0.05$ and $\dfrac{c'}{\gamma H} = 0.025$. As the depth of the circle is not known, the values of r_{ue} in the charts can be used to determine its position.

For $\dfrac{c'}{\gamma H} = 0.05$ and $D_f = 1$ chart (see Fig. 10.6), considering $\beta = 3:1$ and $\phi' = 25°$.

Then $r_{ue} = 0 < r_u$, which indicates that the critical circle passes below this level.

For $\dfrac{c'}{\gamma H} = 0.05$ and $D_f = 1.25$ chart (see Figure 10.6), considering $\beta = 3:1$ and $\phi' = 25°$.

Then $r_{ue} = 0.85 > r_u = 0.15$, which indicates that the critical circle is at this depth.
Reading from the relevant charts, we obtain $m = 2.4$ and $n = 1.8$.

$$F = m - n r_u = 2.4 - 1.8(0.15) = 2.13$$

As there are no broken lines for $\dfrac{c'}{\gamma H} = 0.025$ and $D_f = 1.25$ chart, then the critical circle is not at this level; so considering a lower value of $D_f = 1.0$ chart and considering $\beta = 3:1$ and $\phi' = 25°$, we can obtain $r_{ue} = 0.7 > r_u = 0.15$, which indicates that the critical circle is at this depth.
Reading from the relevant charts, we obtain $m = 1.8$ and $n = 1.65$. Hence:

$$F = m - n r_u = 1.8 - 1.65(0.15) = 1.55$$

Interpolate between the two values to give (Fig. 10.16):

$$\frac{2.13 - F}{2.13 - 1.55} = \frac{0.05 - 0.036}{0.05 - 0.025}$$

$$F = 1.81$$

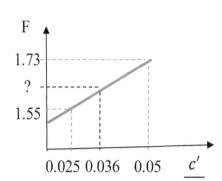

Figure 10.16 Interpolation of FOS.

Worked-out example 10.6: Figure 10.17 indicates the potential instability of a rigid block of rock of weight W under the influence of a water-filled tension crack generating a force V. The base of the block is also subjected to a hydraulic uplift force of U due to water flow through the bedding plane. The base area of the block is A, the cohesion of the infill material of the bedding plane is c and the friction angle is ϕ.

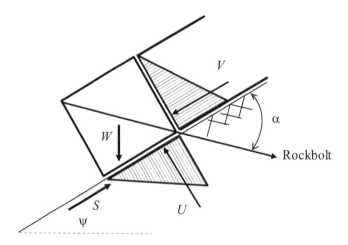

Figure 10.17 Rigid block of rock (modified after Hoek and Bray, 1974).

a.   Determine the factor of safety of the block.
b.   In order to stabilize the block, a prestressed rock anchor is grouted at an inclination α to the slope. What is the minimum anchor force T required to prevent the block from sliding down?

Solution:
a.   The factor of safety can be computed considering the ratio between the resisting and disturbing forces along the direction parallel to the bedding plane, as follows:

$F = (\Sigma \text{ resisting forces})/(\Sigma \text{ disturbing forces})$ or

$$F = \frac{cA + (W\cos\psi - U)\tan\phi}{W\sin\psi + V}$$

b.   If there is a rock bolt, the factor of safety can be given by increasing the numerator; thus:

$$F = \frac{cA + (W\cos\psi - U)\tan\phi + T\sin\alpha}{W\sin\psi + V}$$

Rearranging for T and assuming conditions before failure ($F > 1$),

$$T > \frac{W\sin\psi + V - cA - (W\cos\psi - U)\tan\phi}{\sin\alpha}$$

Worked-out example 10.7: A vertical cut 3.66 m high is excavated in saturated intact clay. The clay is isotropic and has an undrained shear strength (C_u) of 28.7 kN/m² and a saturated unit weight of 18.8 kN/m³. Assuming that a vertical, air-filled tension crack forms behind the slope crest to a depth of 1.53 m, then determine the factor of safety of the cut slope, in terms of shear strength, under short-term (end-of-construction) conditions.

Solution: The problem can be illustrated as shown in Figure 10.18.
   Considering the equilibrium between the resisting and disturbing forces of a finite planar slope, we have:

$$F = \frac{c'l + (W\cos\alpha - Q_w\sin\alpha - P_w)\tan\phi'}{W\sin\alpha + Q_w\cos\alpha}$$

As the tension crack is air filled, then $Q_w = 0$.

Considering no water pressure acting on the slope, i.e. $P_w = 0$.
   For short-term analysis, we have undrained conditions, hence $\phi_u = 0$.

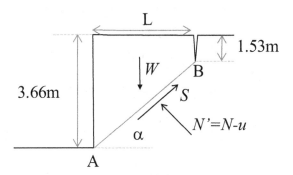

Figure 10.18 Worked-out example 10.7 diagram illustration.

Then the factor of safety can be computed as follows:

$$F = \frac{C_u \times (AB)}{W \sin \alpha}$$

where

$$W = \frac{(3.66 + 1.53)}{2} \times L \times 18.8 = 48.79L$$

$AB = L / \cos \alpha$

Then $F = \dfrac{C_u \times (AB)}{W \sin \alpha} = \dfrac{28.7 \times L / \cos \alpha}{48.79L \sin \alpha} = \dfrac{0.588}{\sin \alpha \cos \alpha}$

As $\sin 2\alpha = 2 \sin\alpha \cos\alpha$ then

$$F = \frac{1.176}{\sin 2\alpha}$$

As $0 < \sin 2\alpha < 1$ then $F_{min} = 1.176 \; for \; \alpha = 45°$

Worked-out example 10.8: An infinite slope exists at an angle β to the horizontal in clay soil having a bulk unit weight of γ and effective strength parameters of c' and ϕ'.

a. Derive the expression of factor of safety against failure along the shallow slip plane parallel to the ground surface.
b. Using the relationship derived, compute the factor of safety for a slope having $c' = 0$, $\phi' = 15°$, $\gamma = 17.5$ kN/m³ and $z = 1.5$, $\beta = 15°$, and assume the water level to be at the ground surface.

Solution:
a. Considering the diagram in Figure 10.19, the shear strength mobilized at the base of the slope is

$\tau = c' + \sigma'_n \tan \phi'; \qquad \sigma'_n = N'/l$

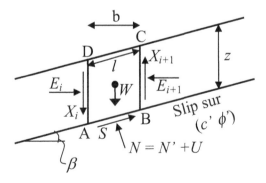

Figure 10.19 Diagram for worked-out problem 10.8.

Resolving forces in the direction parallel to the slip surface gives:

$$W \sin \beta = S = \tau \cdot l \big/ F$$

Considering $W = b\,z\,\gamma = l\,z\,\gamma \cos \beta$,

$$\therefore F = \frac{c' + \left(N'/l\right)\tan\phi'}{\gamma z \sin \beta \cos \beta}$$

Resolving forces in the direction perpendicular to the slip surface gives:

$$N = N' + U = W \cos \beta$$
$$\therefore N' = l\left(\gamma z \cos^2 \beta - u\right)$$

Combining the relationships for directions parallel and perpendicular to the slope, we obtain

$$F = \frac{c' + \left(\gamma z \cos^2 \beta - u\right)\tan\phi'}{\gamma z \sin \beta \cos \beta}$$

b.   Substituting the values given in the previous derived relationship, we obtain:

$$F = \frac{\left(17.5 \times 1.5 \times \cos^2 15 - 1.5 \times 9.81\right)\tan 15°}{17.5 \times 1.5 \times \sin 15° \cos 15°}$$
$$F = \frac{2.62}{6.56}$$
$$F = 0.4$$

References

Bishop, A.W. (1955) The use of the slip circle in the stability analysis of slopes. *Geotechnique*, 5, 7–17.

Bishop, A.W. & Morgenstern, N. (1960) Stability coefficients for earth slopes. *Geotechnique*, 19(4), 129–150.

Fellenius, W. (1936) Calculation of the stability of earth dams. *Transactions 2nd Congress on Large Dams*, 4, 445.

Gibson, R.E. & Morgenstern, N. (1962) A note on the stability of cuttings in normally consolidated clays. *Geotechnique*, 12, 212–216.

Hoek, E. & Bray, J.W. (1974) *Rock Slope Engineering*. The Institute of Mining and Metallurgy, Oxon, United Kingdom, p. 309.

Taylor, D.W. (1937) Stability of earth slopes. *Journal Boston Society of Civil Engineering*, 24, 197–246.

Whitlow, R. (1993) *Basic Soil Mechanics*, 3rd edition, Longman, Harlow, United Kingdom, 536p., ISBN100582236312

Chapter 11

Retaining walls and dams

Determination of lateral earth pressure is essential for the design of retaining structures and dams. The lateral earth pressure acting on a retaining structure is traditionally determined using either the Rankine's or Coulomb's theory of earth pressure. Rankine's theory (1857) considers the stress state in the soil when the shear failure propagates across the entire soil mass. In contrast, Coulomb's theory (1776) considers the stability of the soil wedge between the retaining structure and the potential (trial) failure surface.

11.1 Earth pressure at rest

This is the horizontal soil pressure when the wall is rigid and does not yield; thus the soil is assumed to be in elastic equilibrium. Then the effective horizontal stress is given by: $\sigma'_h = K_0\sigma'_v$, in which K_0 is the coefficient of earth pressure at rest and $\sigma'_v (= \gamma z)$ is the effective vertical stress (Fig. 11.1). Typical values of K_0 are given in Table 11.1.

Jaky (1944) proposed the following equation for a normally consolidated and coarse-grained soil:

$$K_0 = 1 - \sin \phi' \tag{11.1}$$

where ϕ' is the drained friction angle of the soil.

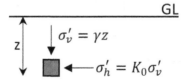

Figure 11.1 Vertical and lateral stress at a point.

Table 11.1 Typical values of K_0 (adapted from Whitlow, 2001).

Soil	K_0
Dense sand	0.45–0.6
Loose sand	0.3–0.5
Normally consolidated clay	0.5–0.7
Overconsolidated clay	1.0–4.0
Compacted clay	0.7–20

11.2 Active earth pressure

Horizontal soil pressure when the wall structure is made to move away from the soil (Fig. 11.2).

11.3 Passive earth pressure

Horizontal soil pressure when the wall structure is moved towards the soil (Fig. 11.2).

11.4 Rankine's theory of earth pressure

This theory considers the ratio of major and minor principal stress when the soil is on the verge of failure throughout its mass, often referred to as plastic equilibrium.

Active pressure for cohesionless soil at any depth z is given by:

$$p_a = K_a \gamma z \tag{11.2}$$

Where p_a is the active pressure, K_a is the active pressure coefficient $\left(K_a = \dfrac{1-\sin\phi'}{1+\sin\phi'} \right)$, and γ is the unit weight of the soil.

The total active thrust (per m length of wall) is given by:

$$P_a = \int_0^H K_a \gamma z dz = \frac{1}{2} K_a \gamma H^2 \tag{11.3}$$

The line of action of P_a passes through the center of the area at a height of 1/3H above the base (Figure.11.3).

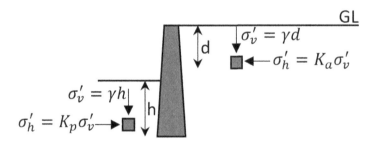

Figure 11.2 Conditions of active and passive earth pressure.

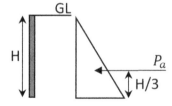

Figure 11.3 Pressure distribution for cohesionless soil (not to scale).

Active pressure for cohesive soil at any depth z is given by:

$$p_a = K_a \gamma z - 2c'\sqrt{K_a} \tag{11.4}$$

Where p_a is active pressure, K_a is the active pressure coefficient, c' is the cohesion and γ is the unit weight of the soil.

For values of $c' > 0$, the value $p_a = 0$ at a particular depth z_c is referred to as the tension crack depth. Substituting $p_a = 0$ in Eq.11.4 gives:

$$\text{Depth of tension crack } z_c = \frac{2c'}{\gamma'\sqrt{K_a}} \tag{11.5}$$

In undrained condition ($\phi = 0$)

$$z_c = \frac{2c_u}{\gamma} \tag{11.6}$$

The total active thrust per unit length of the wall is then given by:

$$P_a = \int_{z_c}^{H} p_a dz = \frac{1}{2} K_a \gamma (H - z_c)^2 \tag{11.7}$$

The force P_a acts at a distance of $1/3$ $(H - z_c)$ above the base (Fig. 11.4).

Passive earth pressure (p_p) for cohesionless soil at any depth z is given by:

$$p_p = K_p \gamma z \tag{11.8}$$

Where p_p is passive pressure, K_p is the passive pressure coefficient $\left(K_p = \dfrac{1+\sin\phi'}{1-\sin\phi'}\right)$, γ is the unit weight of the soil and z is the depth.

The total passive resistance (per m length of wall) is given by:

$$P_p = \int_{0}^{H} K_p \gamma z dz = \frac{1}{2} K_p \gamma H^2 \tag{11.9}$$

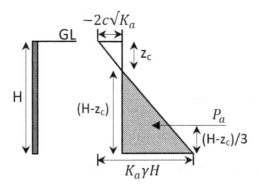

Figure 11.4 Pressure distribution for cohesive soil (not to scale).

The line of action of P_p passes through the center of the area at a height of 1/3H above the base (Fig. 11.5).

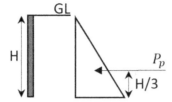

Figure 11.5 Passive pressure distribution for cohesionless soil (not to scale).

Passive pressure for cohesive soil at any depth z is given by:

$$p_p = K_p \gamma z + 2c' \sqrt{K_p} \tag{11.10}$$

where P_p is passive pressure, K_p is the passive pressure coefficient, c' is the cohesion and γ is the unit weight of the soil.

The total passive resistance (per m length of wall) is given by:

$$P_p = \int_0^H \left(K_p \gamma z + 2c' \sqrt{K_p} \right) dz = \frac{1}{2} K_p \gamma H^2 + 2c' \sqrt{K_p} H \tag{11.11}$$

The two components of P_p act a distance of 1/3 H and 1/2 H, respectively, from the base (Fig. 11.6).

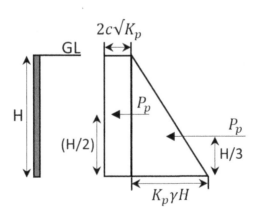

Figure 11.6 Passive pressure distribution for cohesive soil (not to scale).

11.4.1 Effect of water table

For fully drained conditions, the effective unit weight of the soil (γ') and effective stress parameters (c' and ϕ') should be considered for computing active and passive earth pressures.

If the water table is at the surface, then the hydrostatic pressure, $\gamma_w z$, due to the water in the soil pores, must also be considered in addition to the active and passive pressure.

For undrained conditions $(\phi_u = 0)$ in saturated clay, the undrained shear strength (c_u) and the total unit weight (γ_{sat}) should be considered for computing active and passive pressures to check for the immediate stability.

11.4.2 Effect of surcharge load

A retaining wall carrying a uniform surcharge load intensity (q) on top of the backfill may be assumed to cause an equal increase in effective vertical stress along the depth of the wall (Fig. 11.7).

At any depth, z, the additional active earth pressure due to surcharge load is $K_a q$.

At any depth, z, the additional passive earth pressure due to surcharge load is $K_p q$.

11.4.3 Effect of sloping backfill

The active earth pressure is given by:

$$p_a = K_a \gamma z \cos \beta \text{ (acts parallel to the slope)} \tag{11.12}$$

where β is the angle of sloping surface (measured upward) with respect to the horizontal; hence:

$$K_a = \frac{\cos\beta - \sqrt{\cos^2\beta - \cos^2\phi}}{\cos\beta + \sqrt{\cos^2\beta - \cos^2\phi}} \tag{11.13}$$

The passive earth pressure is given by: $p_p = K_p \gamma z \cos\beta$ (acts parallel to the slope)

$$K_p = \frac{\cos\beta + \sqrt{\cos^2\beta - \cos^2\phi}}{\cos\beta - \sqrt{\cos^2\beta - \cos^2\phi}} \tag{11.14}$$

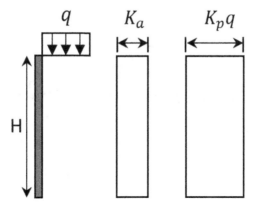

Figure 11.7 Active and passive pressure distribution for surcharge loading (not to scale).

11.5 Coulomb's theory of earth pressure

Coulomb's solution is based on the limit equilibrium consideration of soil wedge between the retaining wall and trial planar failure surface. The force between the wedge and the wall surface can be computed considering the equilibrium forces acting on the soil wedge when it is actively sliding up or down the trial failure surface (Figs. 11.8 and 11.9). Basic theory considers the friction (δ) that had developed between the soil and the retaining wall, thus making the actual failure surface become curved close to the bottom the wall for both active and passive cases. However, the simplified theory assumes a planar failure surface for both active and passive conditions. This assumption has a significant effect on the passive state, especially for $\phi = \delta / 3$.

Active pressure for cohesionless soil at any depth z is given by:

$$p_a = K_a \gamma z \tag{11.15}$$

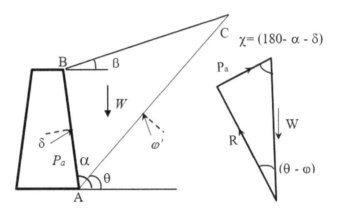

Figure 11.8 Active earth pressure (Coulomb's theory).

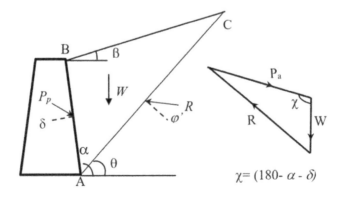

Figure 11.9 Passive earth pressure (Coulomb's theory).

In Eq. 11.15,

$$K_a = \left[\frac{(\sin(\alpha-\phi)/\sin\alpha)}{\sqrt{\sin(\alpha+\delta)} + \sqrt{\left[\dfrac{\sin(\phi+\delta)\cdot\sin(\phi-\beta)}{\sin(\alpha-\beta)}\right]}} \right]^2 \tag{11.16}$$

where δ is the wall friction, α is the inclination of the wall surface with horizontal, β is the inclination of the soil surface with horizontal and ϕ is the friction angle of the soil.

The total active thrust (per m length of wall) is given by:

$$P_p = \int_0^H K_a \gamma z = \frac{1}{2} K_a \gamma H^2 \tag{11.17}$$

Active pressure for cohesive soil at any depth z is given by:

$$p_a = K_a \gamma z - c K_{ac} \tag{11.18}$$

where $K_{ac} = 2\sqrt{K_a\left(1+\dfrac{c_w}{c}\right)}$ and c_w is the wall adhesion parameter.

The depth of tension crack (when $p_a = 0$) can be computed using Eq.11.18.

$$K_a \gamma z - c K_{ac} = 0 \tag{11.19}$$

For undrained condition ($\phi = 0$)

$$z_c = \frac{2c_u\sqrt{\left(1+\dfrac{c_w}{c_u}\right)}}{\gamma} \tag{11.20}$$

Passive earth pressure (p_p) for cohesionless soil at any depth z is given by:

$$p_p = K_p \gamma z \tag{11.21}$$

In Eq. 11.21,

$$K_p = \left[\frac{(\sin(\alpha+\phi)/\sin\alpha)}{\sqrt{\sin(\alpha-\delta)} - \sqrt{\left[\dfrac{\sin(\phi+\delta)\cdot\sin(\phi+\beta)}{\sin(\alpha-\beta)}\right]}} \right]^2 \tag{11.22}$$

The total passive resistance (per m length of wall) is given by:

$$P_p = \int_0^H K_p \gamma z = \frac{1}{2} K_p \gamma H^2 \tag{11.23}$$

Passive pressure for cohesive soil at any depth z is given by:

$$p_p = K_p \gamma z + c K_{pc} \tag{11.24}$$

where $K_{pc} = 2\sqrt{K_p\left(1+\dfrac{c_w}{c}\right)}$.

In the passive case, it is important to consider the curvature of the failure surface to minimize the error, i.e. to avoid overestimating the passive earth pressure. Therefore, it is suggested to use the modified coefficients (K_p) reported by Sokolovski (1965; Fig. 11.10) to determine the passive resistance.

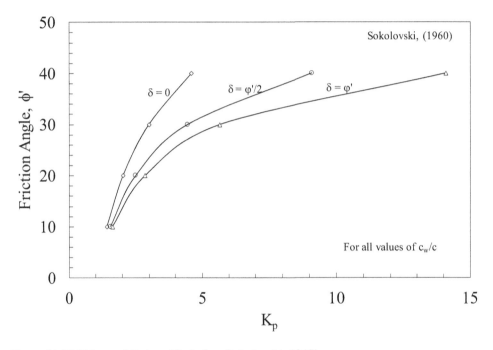

Figure 11.10 Values of K_p (modified after Sokolovski, 1965).

11.6 Stability of gravity walls

Stability of gravity wall is directly related to its self-weight. Sometimes the passive resistance developed in front of the toe of the retaining wall may provide additional stability. In general, design must follow a series of limit state conditions, as follows:

1. Overturning of the wall: the resultant thrust in this case must be within the middle third of the base.

2. The pressure at the base must not exceed the ultimate bearing capacity of the supporting soil. The overturning moment caused by the lateral earth pressure may generate high bearing pressure at the toe of the wall. The maximum and minimum base pressures can then be calculated using:

$$p = \frac{V}{B}\left(1 \pm \frac{6e}{B}\right)$$

where V is the vertical force component; e is the eccentricity and B is width of the base. The eccentricity (e) should not exceed $1/6B$ to ensure the base pressure to remain compressive over the entire base width.

3. Wall sliding: for the sliding limit state, $S = V \tan \delta$ and $S \geq H$, where S is the sliding resistance, H is the horizontal component for force and δ is the friction angle between the base and the underlying soil.

4. Slip surface covers the structure as a whole.

5. Excessive deformation of ground and wall, which may lead to limit state condition.

6. Consideration of seepage, internal erosion and adequacy of drainage system.

7. Failure of structural elements of the wall or combined failure due to faulty design.

Worked-out example 11.1: Determine the total active thrust and passive earth resistance on a vertical smooth retaining wall of height 5 m. The soil properties are: $' = 32°$ and $\gamma = 17$ kN/m³. The water table is well below the base of the retaining wall. (Assume the wall can move towards and away from the retaining wall).

Solution:

$$K_a = \frac{1 - \sin \phi'}{1 + \sin \phi'} = \frac{1 - \sin 32}{1 + \sin 32} = 0.31$$

$$P_a = \frac{1}{2}K_a \gamma H^2 = \frac{1}{2} \times 0.31 \times 17 \times 5^2 = 65.9 \text{ kN/m}$$

$$K_p = \frac{1 + \sin \phi'}{1 - \sin \phi'} = \frac{1 + \sin 32}{1 - \sin 32} = 3.2$$

$$P_p = \frac{1}{2}K_p \gamma H^2 = \frac{1}{2} \times 3.2 \times 17 \times 5^2 = 680 \text{ kN/m}$$

This is illustrated graphically in Figure 11.11.

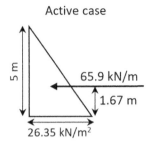

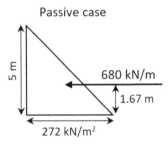

Figure 11.11 Earth pressure diagram of worked-out problem 11.1.

Worked-out example 11.2: A vertical retaining wall supports 10 m of a granular backfill having the following properties: $\phi' = 28°$ and $\gamma = 18$ kN/m³ (Fig. 11.12). The water table is at a depth of 3 m from the horizontal backfill. The saturated unit weight below the water table is $\gamma_{sat} = 19.5$ kN/m³. Determine the total active earth thrust and its point of application.

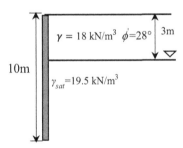

Figure 11.12 Worked-out example 11.2.

Solution:

Consider active pressure distribution:

$$K_a = \frac{1-\sin\phi'}{1+\sin\phi'} = \frac{1-\sin 28°}{1+\sin 28°} = 0.36$$

At $z = 0$

$$p_a = K_a \gamma z = 0$$

At $z = 3$ m,

$$p_a = K_a \gamma z = 0.36 \times 18 \times 3 = 19.44 \text{ kN/m}^2$$

At $z = 10$ m,

$$p_a = K_a \gamma z = 0.36 \times (19.5 - 9.81) \times 10 = 34.9 \text{ kN/m}^2$$

Water pressure distribution:

At $z = 3$ m

$$p_w = \gamma_w z = 0$$

At $z = 10$ m

$$p_w = \gamma_w z = 9.81 \times 7 = 68.67 \text{ kN/m}^2$$

Total active earth thrust is the sum of the active and water pressure distribution diagrams (see Fig.11.13).

$$P_a = 0.5 \times 19.44 \times 3 + 19.44 \times 7 + 0.5 \times 24.46 \times 7 + 0.5 \times 68.67 \times 7 = 491.2 \text{ kN/m}$$

$$\bar{z} = \frac{29.1 \times 8 + 136.1 \times 3.5 + 85.61 \times 2.3 + 240.34 \times 2.3}{491.2} = 2.96 \text{ m}$$

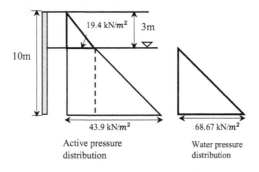

Active pressure distribution

Water pressure distribution

Figure 11.13 Diagram of pressures of worked-out example 11.2.

Worked-out example 11.3: A retaining wall having a smooth vertical back supports a soil as shown in Figure 11.14. Determine the total active thrust acting on the wall.

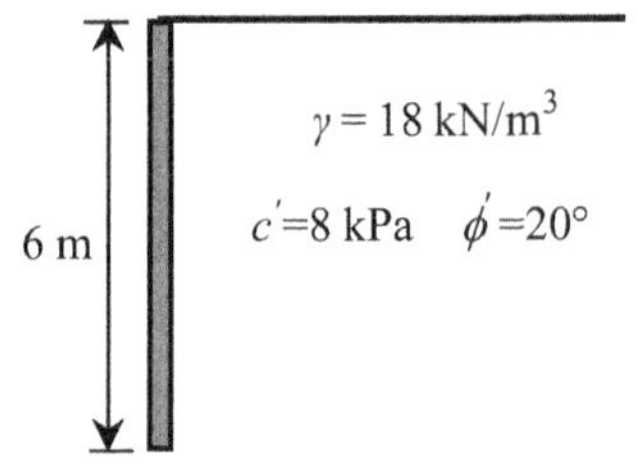

$$\gamma = 18 \text{ kN/m}^3$$
$$c' = 8 \text{ kPa} \quad \phi' = 20°$$

6 m

Figure 11.14 Worked-out example 11.3.

Solution:

$$K_a = \frac{1 - \sin\phi'}{1 + \sin\phi'} = \frac{1 - \sin 20°}{1 + \sin 20°} = 0.5$$

$$z_c = \frac{2c'}{\gamma\sqrt{K_a}} = \frac{2 \times 8}{18 \times \sqrt{0.5}} = 1.26 \text{ m}$$

Considering active pressure distribution (see Fig.11.15):

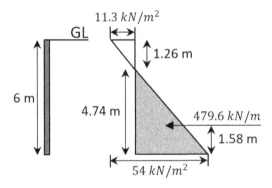

Figure 11.15 Diagram for worked-out example 11.3.

At $z = 0$

$$p_a = K_a\gamma'z - 2c'\sqrt{K_a} = 0 - 2 \times 8 \times \sqrt{0.5} = -11.3 \text{ kN/m}°$$

At $z = 6$ m

$$p_a = K_a\gamma z - 2c'\sqrt{K_a} = 0.5 \times 18 \times 6 - 11.3 = (6 - 1.26) \text{ kN/m}$$

$$P_a = 1/2 \times (6-1.26) \times (54-11.3)^2 = 101.2 \text{ kN/m}$$

$$\bar{z} = 1.58 \text{ m}$$

Worked-out example 11.4: The properties of the soil retained behind a smooth retaining wall are shown in Figure 11.16. Determine the total active earth thrust acting on the back of the wall using a pressure distribution diagram.

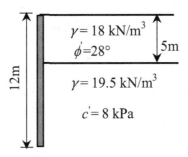

Figure 11.16 Worked-out example 11.4.

Solution:

$$K_a = \frac{1-\sin\phi'}{1+\sin\phi'} = \frac{1-\sin 28°}{1+\sin 28°} = 0.36$$

At $z = 0$

$$p_a = K_a \gamma z = 0$$

At $z = 5$ m

$$p_a = K_a \gamma z = 0.36 \times 18 \times 5 = 32.4 \text{ kN/m}^2$$

At $z = 5$ m

$$p_a = K_a \gamma z - 2c' \sqrt{K_a} = 0.36 \times 18 \times 5 - 2 \times 8 \times \sqrt{0.36} = 22.8 \text{ kN/m}^2$$

At $z = 12$ m

$$p_a = K_a \gamma z - 2c' \sqrt{K_a} = 0.36(18 \times 5 + 7 \times 19.5) - 2 \times 8 \times \sqrt{0.36} = 71.94 \text{ kN/m}^2$$

Total active thrust (see Fig. 11.17):

$$P_a = \frac{1}{2} \times 32.4 \times 5 + 22.8 \times 7 + \frac{1}{2} \times 49.14 \times 7 = 412.6 \text{ kN/m}$$

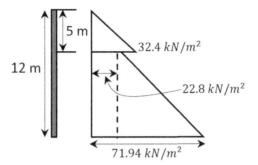

Figure 11.17 Diagram for worked-out example 11.4.

Worked-out example 11.5: Determine the total active earth thrust on a vertical retaining wall of height 6 m (Fig.11.18). A uniform surcharge of 50 kPa is place on the surface of the horizontal backfill. The soil properties are: $\phi' = 30°$, $\gamma = 18$ kN/m³ and $c' = 0$.

Solution:

$$K_a = \frac{1-\sin\phi'}{1+\sin\phi'} = \frac{1-\sin 30°}{1+\sin 30°} = 1/3$$

At $z = 6$ m

$$p_a = K_a\gamma z = \frac{1}{3}\times 18\times 6 = 36 \text{ kN/m}^2$$

Active thrust due to surcharge load (Fig.11.18) is determined by:

$$p_a = K_a q = \frac{1}{3}\times 50 = 16.67 \text{ kN/m}^2$$

Total active thrust is now given by:

$$P_a = \frac{1}{2}\times 36\times 6 + 16.67\times 6 = 208 \text{ kN/m}$$

$$\bar{z} = \frac{108\times 2 + 100\times 3}{208} = 2.5 \text{ m}$$

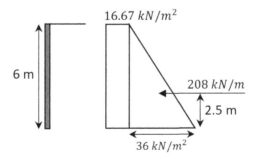

Figure 11.18 Diagram for worked-out example 11.5.

Worked-out example 11.6: Check the design of the gravity retaining wall shown in Figure 11.19. The bearing pressure of the foundation soil should not exceed 450 kPa. Assume unit weight of the wall = 23.5 kN/m³.

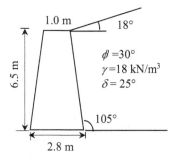

Assume unit weight of the wall, = 23.5 kN/m³

1.0 m   18°

ϕ =30°
γ=18 kN/m³
δ= 25°

6.5 m

105°

2.8 m

Figure 11.19 Worked-out example 11.6.

Solution: Coulomb's theory is used to compute the active earth pressure considering the wall friction and the inclination of wall and soil surfaces (see Section 11.5).

$$K_a = \left[\frac{(\sin(\alpha - \phi)/\sin\alpha)}{\sqrt{\sin(\alpha + \delta)} + \sqrt{\left[\frac{\sin(\phi + \delta)\cdot\sin(\phi - \beta)}{\sin(\alpha - \beta)} \right]}} \right]^2 = 0.6$$

From Eq.11.17:

$$P_a = \frac{1}{2}K_a\gamma H^2 = \frac{1}{2}\times0.6\times18\times6.5^2 = 228.15\,\text{kN/m}^2$$ and it acts at 40° above the horizontal.

Table 11.2 shows the moment of all forces about the toe of the gravity wall.

Table 11.2 Calculations for worked-out example 11.6.

Pressures (/m)	Force (kN)	Lever arm (m)	Moment (kNm)
Wall	$\frac{1}{2}\times0.9\times6.5\times23.5 =$	0.6	41.25
		1.4	214.2
		2.2	151.25
	68.75		
	$1\times6.5\times23.5 = 153$		
	$\frac{1}{2}\times0.9\times6.5\times23.5 =$		
	68.75		
$P_a \sin40°$	147	2.2	323.4
	V = 438		M_v = 730
$P_a \cos40°$	175	2.2	385
	H = 175		M_H = 385
			$\Sigma M = M_v - M_H = 345$

Lever arm of base resultant: $\sum \dfrac{M}{V} = \dfrac{345}{438} = 0.8\,\text{m}$

Check for overturning

The restoring moment (M_v) is greater than the disturbing moment (M_H). Hence this satisfies the overturning limit state.

Check for bearing pressure

Eccentricity of base reaction, $e = \dfrac{2.8}{2} - 0.8 = 0.6\,\text{m}$

$$p_{max} = \dfrac{V}{B}\left(1 + \dfrac{6e}{B}\right) = \dfrac{438}{2.8}\left(1 + \dfrac{6 \times 0.6}{2.8}\right) = 357.5\,\text{kN/m}^2$$

The maximum bearing pressure is less than the allowable bearing pressure of the foundation soil, i.e. 450 kPa. This satisfies the bearing pressure limit state.

Check for sliding

The restoring force:

$\quad V \tan \delta = 438 \times \tan 25° = 204.2\ \text{kN}$

The disturbing force:

$\quad H = 185\ \text{kN}.$

Hence this satisfies the sliding limit state.
Therefore, the design of the gravity wall is satisfactory.

References

Coulomb, C.A. (1776) Essai sur une application des regles de maximis et minimis a quelqes proble-mas de stratique relatifs a l'architecture. *Memoires de mathematique et de physique. Presentes a l'Academie Royale des Sciences*, Paris, vol. 7, pp. 343–382.

Jaky, J. (1944) The coefficient of earth pressure at rest (in Hungarian). *Journal of the Society of Hungarian Architect and Engineers*, 78(22), 355–358.

Rankine, W.J.H. (1857) On the stability of loose earth. *Philosophical Transactions of the Royal Society*, 147.

Sokolovski, V.V. (1965) *Statics of Granular Media*. Pergamon Press, Oxford.

Whitlow, R. (2001) *Basic Soil Mechanics*, Pearson Prentice Hall, New York.

Chapter 12

Excavations and tunnels

This chapter introduces different approaches for estimating the extent of ground movement and the degree of stability of vertical cuttings, trenches and tunnels as essential in the development of civil infrastructure.

12.1 Vertical cuttings and trenches

For the excavation of trial pits and trenches, undrained stability analysis considering total stresses is often considered in traditional design approaches. Limit equilibrium can be conveniently adopted to assess the stability of a vertical cut.

Considering the soil wedge shown in Figure 12.1 of a vertical cutting, the equilibrium of forces in the horizontal and vertical directions can be resolved as follows:

$$S \cos \beta - N \sin \beta = 0 \tag{12.1a}$$

$$W - S \sin \beta - N \cos \beta = 0 \tag{12.1b}$$

The weight of the wedge (W) and shear force (S) acting on the translational failure plane (dashed line) can be derived by limit equilibrium condition, thus:

$$W = \frac{\gamma H^2}{2 \tan \beta} \tag{12.2}$$

$$S = \frac{c_u H}{\sin \beta} \tag{12.3}$$

In Eq. 12.2, γ is the unit weight of the soil and c_u is the undrained shear strength. Note that $H / \sin \beta$ corresponds to the area of the slip plane per meter length.

Rearranging the terms in Eq. 12.1a gives:

$$N = S \frac{\cos \beta}{\sin \beta} \tag{12.4}$$

Substituting W, S and N into Eq. 12.1b and rearranging for H gives:

$$H = \frac{2c_u}{\gamma \sin \beta \cos\beta} \tag{12.5}$$

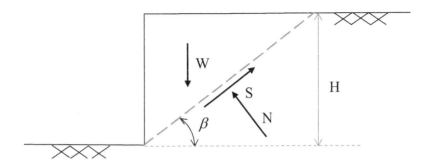

Figure 12.1 Vertical cutting diagram.

The value of H computed in Eq.12.5 is the maximum height of the vertical cut such that limit equilibrium is maintained; H is an upper bound given the maximum undrained shear strength (c_u), which in practice is not always fully mobilized. As in Chapter 10, the consideration of the factor of safety in evaluating stability of the vertical cutting and excavations is necessary in design practice.

Also noteworthy, there are drained conditions to be considered where appropriate. In such cases, the steps of analysis remain the same, but the shear resistance (S) term needs to be modified to capture the *effective* (drained) stresses as follows:

$$S = c'\left(\frac{H}{\sin \beta}\right) + (N - U)\tan \phi' \tag{12.6}$$

where c' and ϕ' are the effective shear strength parameters, cohesion and friction angle, respectively, and U is the hydraulic uplift force acting on the plane of sliding, thus:

$$U = \frac{1}{2}\gamma_w h_w^2 \text{cosec}\beta \tag{12.7}$$

Where h_w is the height of the groundwater table measured from the bottom of excavation level and γ_w is the unit weight of the water.

12.2 Tunnels

The method of tunnel construction depends on several factors such as the ground profile, the groundwater level, the depth and dimensions of the tunnel, and the final geometric shape of the tunnel (oval, horseshoe or circular). There are three main groups of tunnels: (i) cut-and-cover, (ii) bored and (iii) immersed tunnels.

Cut-and-cover tunnels are usually constructed in a shallow trench, and the stability requirements follow typically those of open excavations. For bored tunnels, there are two main types of construction methods generally adopted, which include open-face tunneling that is optimal for noncircular sections and flexible construction sequencing and closed-face tunneling methods which typically involve the use of a tunnel-boring machine (TBM).

12.2.1 Bored tunnels

12.2.1.1 Stability, ground movement and tunnel linings

The assessment of tunnel face stability is critical in urban environments mainly because of possible catastrophic consequences (e.g. sinkholes appearing in the Brightwater Conveyance tunnel in the U.S. or caving in the Cologne metro tunnel in Germany). The stability of the opening prior to the installation of the tunnel lining, hydraulic pressures and permeability of the surrounding ground, rate of excavation (e.g. unsupported face length, P), size of the tunnel (e.g. diameter, D) and tunnel depth (C) are the most important factors considered in analysis (Fig. 12.2).

In soft-ground tunneling, based on force equilibrium at the tunnel heading, the stability factor N takes into consideration the total applied stress in relation to the undrained shear strength of the excavated soil. Adopting the concept introduced by Broms and Bennermark (1967), the stability ratio N is then defined by:

$$N = \frac{\sigma_s + \gamma z_0 - \sigma_T}{c_u} \tag{12.8}$$

where γ = unit weight of the soil (kN/m³), z_0 = depth of the tunnel axis ($C + D/2$), σ_s = surface surcharge pressure, σ_T = tunnel support pressure and c_u = undrained shear strength at the tunnel axis level.

Field observations (Peck, 1969) have shown that the value of N between 5 and 7 may indicate instability at the tunnel face. More recently, based on centrifuge testing,

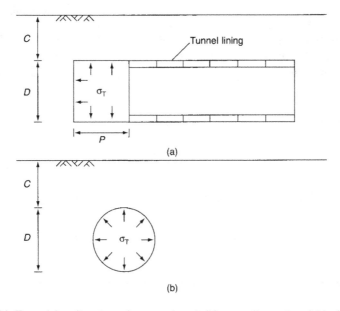

Figure 12.2 (a) Tunnel heading in soft ground and (b) two dimensional idealization (after Mair and Taylor, 1997, reproduced with permission from CRC Press).

a range of N values has been established as follows (ITA-AITES WG 'Research', 2007):

- $N \leq 3$ implies that the stability of the tunnel face is ensured
- $3 < N \leq 6$ suggests that surface settlement may be significant, as excessive ground losses are expected for $N > 5$
- $N > 6$ implies that the tunnel face is unstable and the ground losses at the face are unacceptable

A critical stability ratio or N_c may be considered to evaluate the short-term stability of tunnels. N_c can be inferred from Figures 12.3a and b. Note that if $N < N_c$, then the tunnel is stable.

In relation to Figure 12.3a, for $P/D = 0$, the stability chart can be further extended (Fig. 12.3b). This approach may also be used to estimate the risk of "blowout," which can occur if the tunnel face pressure is too high particularly in soft soils.

The relationship between surface settlement and tunnel depth is neither simple nor linear. In reality, ground movements depend on a number of factors including (i) geological, hydrogeological and geotechnical conditions, (ii) tunnel geometry and depth, (iii) excavation methods and (iv) the quality of workmanship and management. It is, however, clear that a shallow tunnel will tend to have a greater effect on surface structures than a relatively deep one.

The surface and subsurface movements are directly related to the ground loss, which is the volume of soil that exceeded the theoretical volume of excavation, normally expressed as a percentage of the tunnel area per meter length. The main components of ground loss can be categorized into five quantifiable entities as follows (Fig. 12.4):

1. Face and heading effect, S_f and S_h: deformation of the ground towards the opening resulting from stress relief.
2. Over-cut shield loss, S_s: The presence of an overcutting edge (bead) combined with any tendency of the machine to plough or yaw will lead to radial ground movements.
3. Tail loss (soil lining), S_t: The existence of a gap between the skin of the shield and the lining in the tail section will cause some radial ground movements.
4. Deformation of the lining due to earth pressure, S_l: deflection of the lining as the ground loading develops.

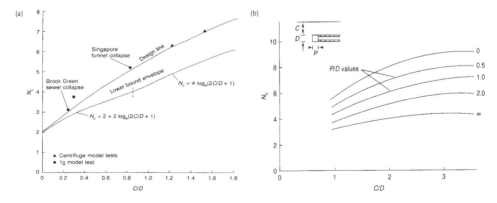

Figure 12.3 Dependence of the critical stability ratio on tunnel heading geometry (after Mair and Taylor, 1997, reproduced with permission from CRC Press).

For excavations without a shield, components 2 and 3 are not accounted. Besides the ground loss, a fifth component must be considered for predicting total ground movements:

5. Soil volumetric change, S_c: Consolidation associated with the long-term dissipation of pore water pressures and dilation/contraction due to tunneling-induced shear stresses.

The total volume loss (V_L) is the sum of all five components (Fig. 12.4):

$$V_L = S_f + S_h + S_s + S_t + S_l + S_c \tag{12.9}$$

The vertical subsidence caused by the total volume loss can be represented by a transverse settlement profile, which is based on field observations, and can be described by a Gaussian distribution curve as suggested by Peck (1969) as follows:

$$S_v(y) = S_{max}\exp\left(-\frac{y^2}{2i^2}\right) \tag{12.10}$$

Where $S_v(y)$ = settlement at any given y, i = trough width parameter, S_{max} = maximum vertical settlement and y = distance from the tunnel centerline.

Marshall et al. (2012) compared the performance of different empirical equations of transverse settlement, and Eq. 12.10 could predict the observed behavior reasonably well; however, models with greater number of degrees of freedom (e.g. Vorster et al., 2005) showed superior performance for settlement prediction in tunnels in sand.

The volume of surface settlement, V_s (per meter length of tunnel), can be evaluated by integrating Eq. 12.10 to give:

$$V_s = \sqrt{2\pi}\, i\, S_{max} \tag{12.11}$$

The trough width parameter (i) may be estimated by adopting empirical relationships available from previous studies. Table 12.1 summarizes the relationships proposed in past studies based on field observations and limited centrifuge data.

Typically, the volume of surface settlement (V_s) can be related to the total volume loss (V_L), according to the type of material and ground conditions:

$$V_s = f(V_L) \tag{12.12}$$

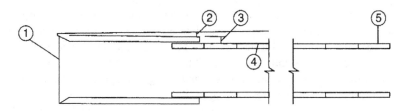

Figure 12.4 Primary components of ground movement (after Mair and Taylor, 1997, reproduced with permission from CRC Press).

Table 12.1 Empirical relationships proposed for the estimation of *i*.

Width through parameter (i) value relationship	Reference
$i = 0.25(z_0 + R)$; loose sand	Atkinson and Potts (1979)
$i = 0.25(1.5z_0 + 0.5R)$; dense sand/overconsolidated clay	
$i = 0.43z_0 + 1.1$; cohesive soil	O'Reilly and New (1982)
$i = 0.28z_0 - 0.1$; granular soil	
$i = 0.5z_0$	Mair et al. (1996)
	Clough and Schmidt (1981)
$\dfrac{i}{R} = \alpha \left(\dfrac{z_0}{2R}\right)^n$; $\alpha = 1$ and $n = 0.8$	

Note: z_0 is the depth of the tunnel axis below ground level (i.e. at the springline) and R is the tunnel radius.

As a general guide (Atkinson and Potts 1979):

- $V_s = 0$ (very deep tunnels)
- $V_L > V_s$ for dense sands (drained conditions) because of dilation
- $V_L \approx V_s$ for undrained saturated clay, because typically in undrained conditions ground movement will occur under constant volume
- $V_L < V_s$ for loose sands because they exhibit mainly contractive behavior

Knowing the equivalent excavated tunnel volume in soil unit of volume or V_0, the total volume can be computed based on an empirical relationship proposed by Dimmock and Mair (2007):

$$\frac{V_L}{V_0} = 0.23e^{4.8 \times LF} \,(\%) \tag{12.13}$$

where V_0 = equivalent excavated tunnel soil unit of volume and LF is the ratio between N and N_c.

12.2.1.2 Stresses at the tunnel level: Terzaghi's arching theory (Terzaghi, 1943)

Terzaghi (1943) proposed a theoretical approach considering soil arching under plane strain conditions. Figure 12.5a shows a section through a bed of sand in which a tunnel is to be constructed between the horizontal surface of the bed and the water table. Due to the imperfect fit of the support connections and the compressibility of its footings, the support deformation is usually sufficient to reduce the stress in the sand (i.e. ground pressure acting onto the tunnel lining) almost to the value corresponding to the shear failure state. This is similar to the stress state in a mass of sand above a yielding strip. The sand adjoining the sides of the tunnel also subsides due to the deformation of its lateral supports.

Based on traditional earth pressure theory, the inclined boundaries of the zone of subsidence rise at an angle of about $45° - \phi/2$ (Fig. 12.5a). Therefore, at the level of the tunnel roof, the width of the yielding strip is approximately equal to:

$$2B_1 = 2\left[B_0 + H \tan\left(45 + \frac{\phi}{2} \right) \right]$$

(12.14)

If a tunnel is located at a greater depth below the surface, the arching effect does not extend beyond a certain elevation D_1 above the tunnel roof. The soil located above this elevation, from the surface of the ground down to a depth D_2 (Fig. 12.5b), acts on the zone of arching like a simple surcharge with an intensity γD_2 per unit area. In this case, the pressure on the tunnel roof is determined by Eq.12.15.

$$\sigma_v = \frac{\gamma B_1}{k_0 \tan \phi}\left(1 - e^{\left(-k_0 \tan \phi \; {}^{D_1}\!/_{B_1}\right)} \right) + \gamma D_2 e^{\left(-k_0 \tan \phi \; {}^{D_1}\!/_{B_1}\right)}$$

(12.15)

If the roof of a deep tunnel yields, the height D_1 of the zone of arching increases while the height D_2 decreases. The first term of Eq. 12.15 is smaller than $\gamma B_1/K_0 \tan \phi$ for all values of D_1. Hence, the pressure per unit of area of a deep tunnel through soil does not exceed an upper limiting value, as given by:

$$\sigma_{v,0} = \frac{\gamma B_1}{k_0 \tan \phi}$$

(12.16)

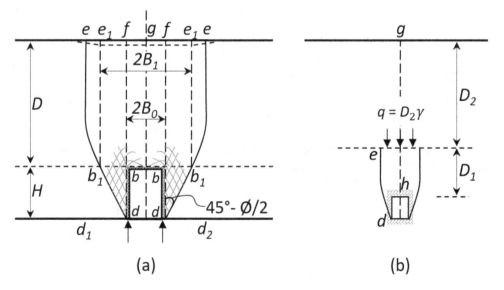

Figure 12.5 (a) Flow of soil toward a shallow tunnel when yielding happened in the soil body and (b) vertical stress profile in the soil located above the tunnel for a deep tunnel (modified after Terzaghi, 1943).

12.3 Immersed tube tunnels

Immersed tube tunnels are sunk into a body of water and sit on or are buried just under its bed. Two main conditions need to be analyzed while considering immersed tunnels (Fig. 12.6), during transport (flotation) and sinking into position.

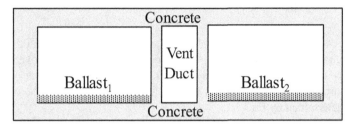

Figure 12.6 Schematic illustration of the different elements contributing to the weight balance during transport and installation of an immersed tunnel.

Considering force equilibrium, at the transport stage:

$$W_{\text{tunnel, concrete}} = W_{\text{water, displaced}}$$
$$\gamma_c \times V_{\text{tunnel, concrete}} = \gamma_w \times V_{\text{tunnel, outline}}$$

(12.17)

where γ_c is the concrete unit weight and γ_w is the water unit weight.

In the final phase, when the tunnel is sunk into position, a counter flotation margin (f_m) should apply; hence:

$$W_{\text{tunnel, concrete}} + W_{\text{ballast}} = W_{\text{water, displaced}} \times f_m$$
$$\gamma_c \times V_{\text{tunnel, concrete}} + \gamma_b \times V_{\text{ballast}} = \gamma_w \times V_{\text{tunnel, outline}} \times f_m$$

(12.18)

where γ_b is the ballast unit weight.

Considering these concepts, theories and summary notes, the following worked-out examples are now presented.

Worked-out example 12.1: A vertical cutting in a clay deposit having an undrained shear strength c_u of 20 kPa is proposed for a road project (Fig. 12.7).

a.   Derive the relationship for the maximum height of the cutting based on force equilibrium.
b.   Calculate the maximum height for a vertical cutting on undrained clay having c_u of 25 kPa and unit weight of 17 kN/m³.

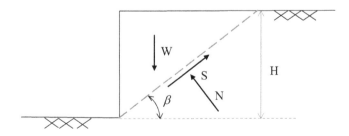

Figure 12.7 Vertical cutting diagram for workout example 12.1.

Solution:

a.   Derive $H = \dfrac{2c_u}{\gamma \sin \beta \cos \beta}$ based on force equilibrium for a soil wedge considered in Figure 12.7.

The force equilibrium in the vertical and horizontal directions are as follows:

$$S\cos\beta - N\sin\beta = 0$$
$$W - S\sin\beta - N\cos\beta = 0$$

where

$$W = \frac{\gamma H^2}{2\tan\beta}$$

$$S = c_u H / \sin\beta$$

$$N = S\frac{\cos\beta}{\sin\beta} \quad \text{(force equilibrium in horizontal direction)}$$

Substituting W, S and N into Eq. 12.1b, then

$$\frac{\gamma H^2}{2\tan\beta} - c_u\frac{H}{\sin\beta}\sin\beta - c_u\frac{H}{\sin\beta}\frac{\cos\beta}{\sin\beta}\cos\beta = 0$$

$$\frac{\gamma H^2}{2\tan\beta} - c_u H - c_u\frac{H}{\tan^2\beta} = 0$$

$$\frac{\gamma H}{2} - c_u\tan\beta - \frac{c_u}{\tan\beta} = 0$$

$$H = \left(\frac{1}{\tan\beta} + \tan\beta\right)c_u\frac{2}{\gamma}$$

$$H = \left(\sin^2\beta + \cos^2\beta\right)\frac{1}{\sin\beta\,\cos\beta}c_u\frac{2}{\gamma}$$

$$H = \frac{2c_u}{\gamma \sin\,\beta\cos\beta}$$

b.   The cutting will fail when H is minimum. The value of β at which the minimum H occurs (wedge failure) may be found by solving $dH/d\beta = 0$ or $\beta = \pi/4 = 45°$ and then substituting in the expression for H.

$$H = \frac{2c_u}{\gamma \sin\,\beta\cos\beta} \quad \text{or} \quad H = \frac{2\times 25}{17\times\sin 45\cos 45} = 5.88\,\text{m}$$

Thus, the height that can be excavated for this cutting should not exceed 5.88 m for stability.

Worked-out example 12.2: A trench for installing a stormwater pipe is proposed in a saturated clay site having c_u of 25 kPa and unit weight of 17 kN/m³. The unit weight of the bentonite slurry employed to assist in supporting the excavation cutting has 11.5 kN/m³ (Fig. 12.8).

a. Derive the relationship for the maximum height of the trench based on force equilibrium considering undrained conditions (e.g. short-term analysis).
b. Calculate the maximum height of the trench if the trench is filled with slurry to depth of 80% of the total height of the excavation (i.e. $x = 0.8$).

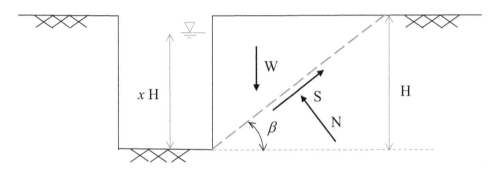

Figure 12.8 Diagram illustrating a slurry-supported trench in undrained soil.

Solution:
a. Considering force equilibrium for a soil wedge shown in Figure 12.8:

$$P + S\cos\beta - N\sin\beta = 0$$
$$W - S\sin\beta - N\cos\beta = 0$$

where

$$W = \frac{\gamma H^2}{2\tan\beta}$$

$$S = \frac{c_u H}{\sin\beta}$$

$$N = S\frac{\cos\beta}{\sin\beta} \quad \text{(force equilibrium in horizontal direction, empty trench)}$$

$$P = \int_0^{0.9H} \gamma_s z\, dz = \frac{\gamma_s (xH)^2}{2} \quad \text{(hydrostatic pressure distribution in the trench derived from the slurry)}$$

As $P + S\cos\beta - N\sin\beta = 0$, then $N = \dfrac{P + S\cos\beta}{\sin\beta}$

Substituting W, S and N into Eq. 12.1b,

$$\frac{\gamma H^2}{2\tan\beta} - c_u\frac{H}{\sin\beta}\sin\beta - \frac{P+S\cos\beta}{\sin\beta}\cos\beta = 0 \ \text{ or}$$

$$\frac{\gamma H^2}{2\tan\beta} - c_u H - \left(\frac{\gamma_s(xH)^2}{2} + \frac{c_u H\cos\beta}{\sin\beta}\right)\frac{\cos\beta}{\sin\beta} = 0$$

$$\frac{\gamma H^2}{2\tan\beta} - c_u H - \frac{\gamma_s(xH)^2}{2\tan\beta} - c_u\frac{H}{\tan^2\beta} = 0$$

Multiplying all terms by $\dfrac{\tan\beta}{H}$, then

$$\frac{\gamma H}{2} - c_u\tan\beta - \frac{\gamma_s x^2 H}{2} - \frac{c_u}{\tan\beta} = 0$$

$$H\frac{(\gamma - \gamma_s x^2)}{2} = \frac{c_u}{\tan\beta} + C_u\tan\beta$$

$$H = \frac{2c_u}{\sin\beta\cos\beta}\left[\frac{1}{(\gamma - \gamma_s x^2)}\right]$$

or

$$H = \frac{2c_u}{\gamma\sin\beta\cos\beta}\left[\frac{1}{1 - \left(\dfrac{\gamma_s}{\gamma}\right)x^2}\right]$$

Note that: $\left(\dfrac{1}{\tan\beta} + \tan\beta\right) = (\sin^2\beta + \cos^2\beta)\dfrac{1}{\sin\beta\cos\beta} = \dfrac{1}{\sin\beta\cos\beta}$

b. Substituting the values for the problem and assuming near-failure conditions for which β approaches $45°$, then

$$H = \frac{2c_u}{\gamma\sin\beta\cos\beta}\left[\frac{1}{1 - \left(\dfrac{\gamma_s}{\gamma}\right)x^2}\right] \ \text{ or}$$

$$H = \frac{2\times 25}{17\sin 45°\cos 45°}\left[\frac{1}{1 - \left(\dfrac{11.5}{17}\right)(0.8)^2}\right] = 5.88\times 1.76 = 10.35\,\text{m}$$

The maximum depth of the trench is 10.35 m.

The consideration of slurry support enables the excavation depth to be increased by approximately 76% without compromising stability.

Worked-out example 12.3: A new cut-and-cover tunnel is proposed for a new subway line. The tunnel is to be constructed in precast concrete with an average depth of cover of 3 m. The preliminary site investigation and laboratory testing indicated the presence of a clay deposit (γ_{sat} = 18.5 kN/m³, c_u = 30 kPa, K_0 = 0.8) and detected the groundwater level at 1 m depth. Assuming the ground profile is relatively homogeneous, calculate the stresses acting on the tunnel during service for a design surcharge of 35 kPa.

Solution: During construction, the water level may be lowered, and the stability of the excavation may be assessed using the vertical cutting theory.

For long-term assessments, the groundwater level is restored to its original position, and the stresses acting on the precast concrete box are illustrated as shown in Figure 12.9.

Assuming that the material used to fill the trench has the same properties of the original clay, the stresses acting on the top of the tunnel can be calculated considering the overburden and surcharge pressures and the restored groundwater level, thus:

$\sigma_v = 3 \times 18.5 + 35 = 90.5$ kPa
$P_w = 2 \times 9.81 = 19.62$ kPa
$\sigma'_v = 90.5 - 19.62 = 70.88$ kPa
$\sigma'_h = 0.8 \times 70.88 = 62.37$ kPa

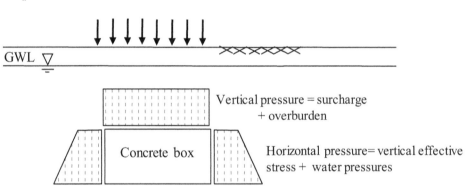

Figure 12.9 Schematic diagram of the stresses acting on a precast concrete tunnel constructed using cut-cover method.

Worked-out example 12.4: A tunnel having a diameter of 5 m is to be constructed with 10 m cover in a clay deposit. Site investigations revealed that the groundwater level was located at the surface and the clay deposit has γ_{sat} = 20 kN/m³, c_u = 25 kPa, E = 20 MPa, K_0 = 0.5, υ = 0.3.

a.   Determine the minimum tunnel support pressure to ensure stability (fully supported face),
b.   the maximum settlement at surface caused by the excavation of the tunnel and
c.   transverse surface settlement profile.

Solution: The profile of the ground can be represented by Figure 12.10.

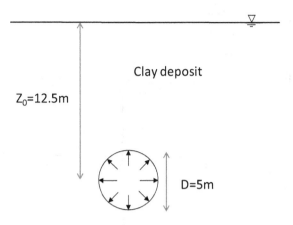

Figure 12.10 Ground profile of worked-out example 12.4.

a. To ensure stability conditions $N \leq 3$, then from Eq. 12.8:

$$N = \frac{\sigma_s + \gamma z - \sigma_T}{C_u} \, or \, \frac{0 + 20 \times 12.5 - \sigma_T}{25} \leq 3$$

$\sigma_T \geq 175 \, \text{kPa}$

b. As it is a circular tunnel, the volume occupied per unit meter length of the tunnel is given by:

$$V_0 = \pi r^2 \times 1 = \pi \times 2.5^2 \times 1 = 19.63 \, \text{m}^3$$

To calculate the total volume loss (V_L), we may use Dimmock and Mair's (2007) approach for overconsolidated clays; hence,

$$\frac{V_L}{V_0} = 0.23e^{4.8 \times LF} \, \text{ and } \, LF = N/N_c$$

To ensure stability at the face, $N \leq 3$ and N_c may be obtained considering Figure 12.11, represented as follows:

For $C/D = 2$ and $P/D = 0$, N_c is 8, then $LF \geq 3/8$ or $LF \geq 0.375$.

$$\frac{V_L}{V_0} = \frac{0.23e^{4.8 \times LF}}{100}$$

$$\frac{V_L}{19.63} = \frac{0.23e^{4.8 \times 0.375}}{100}$$

$V_L = 0.27 \, \text{m}^3$ lengh of tunnel

For undrained clay, assuming short-term analysis during tunnel excavation, $V_L \cong V_s$, we obtain:

$$V_s = \sqrt{2\pi} \, i \, S_{max} \, \text{ or } \, S_{max} = \frac{V_s}{2.5i}$$

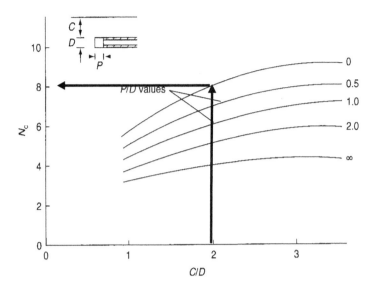

Figure 12.11 Dependence of the critical stability ratio on tunnel heading geometry (after Mair and Taylor, 1997, reproduced with permission from CRC Press).

Thus, to calculate the maximum surface settlement, we need to compute the trough width parameter i as follows:

- Atkinson and Potts (1979): $i = 0.25(1.5z_0 + 0.5R)$ or $i = 0.25(18.75 + 1.25) = 5.00$
- O'Reilly and New (1982): $i = 0.43z_0 + 1.1 = 0.43 \times 12.5 + 1.1 = 6.48$
- Clough and Schmidt (1981): $i = \left(\dfrac{D}{2}\right)\left(\dfrac{z_0}{D}\right)^{0.8} = 5.2$

The maximum surface settlement caused by the tunnel excavation ranges from 8.64 to 21.6 mm.

c. For illustrating the transverse surface settlement profile, the relationship proposed by Peck (1969) may be adopted using the Gaussian distribution:

$$S_v(y) = S_{max}\exp\left(-\frac{y^2}{2i^2}\right)$$

Then the profile can be evaluated between $\pm 2.5i$.

Table 12.2 shows the values computed for the different trough width parameter, and for the values determined earlier we obtain the following (Table 12.3, Fig. 12.12):

Table 12.2 Maximum surface settlement for different values of i.

$S_{max} = \dfrac{V_s}{2.5i}$	
21.6 mm	Atkinson and Potts (1979)
16.7 mm	O'Reilly and New (1982)
20.8 mm	Clough and Schmidt (1981)

Table 12.3 Values computed for the transverse surface settlement profile.

Y	S(y), i = 5	S(y), i = 6.48	S(y), i = 5.2
−30	3.29E-10	3.68E-07	1.23E-09
−27	1.01E-08	2.82E-06	2.91E-08
−24	2.14E-07	1.74E-05	4.93E-07
−21	3.19E-06	8.70E-05	5.98E-06
−18	3.31E-05	3.50E-04	5.20E-05
−15	2.40E-04	1.14E-03	3.24E-04
−12	1.21E-03	2.99E-03	1.45E-03
−9	4.27E-03	6.33E-03	4.65E-03
−6	1.05E-02	1.08E-02	1.07E-02
−3	1.80E-02	1.49E-02	1.76E-02
0	2.16E-02	1.66E-02	2.08E-02
3	1.80E-02	1.49E-02	1.76E-02
6	1.05E-02	1.08E-02	1.07E-02
9	4.27E-03	6.33E-03	4.65E-03
12	1.21E-03	2.99E-03	1.45E-03
15	2.40E-04	1.14E-03	3.24E-04
18	3.31E-05	3.50E-04	5.20E-05
21	3.19E-06	8.70E-05	5.98E-06
24	2.14E-07	1.74E-05	4.93E-07
27	1.01E-08	2.82E-06	2.91E-08
30	3.29E-10	3.68E-07	1.23E-09

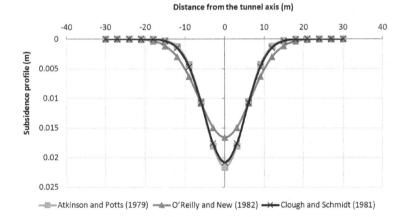

Figure 12.12 Transversal surface settlement profile of worked-out example 12.4.

Worked-out example 12.5: For the tunnel considered in worked-out example 12.4,

a. calculate the tunnel face pressure to ensure that surface settlement of 10 mm is not excessed, and

b. if there is a surface surcharge of approximately 75 kPa (e.g. buildings).

Solution:

a. For this case, adopt the most conservative estimate for surface settlement, i.e. the estimation of i provided by Atkinson and Potts (1979), then:

$$S_{max} = \frac{V_s}{2.5i} \text{ or } V_s = S_{max} \times 2.5i$$

$$V_s = 0.01 \times 2.5 \times 5 = \frac{0.125m^3}{m}$$

Substituting the relevant values,

$$\frac{V_L}{V_0} = \frac{0.23e^{4.8 \times LF}}{100}$$

$$\frac{0.125}{19.63} = \frac{0.23e^{4.8 \times LF}}{100}$$

$$LF = 0.21$$

As N_c is 8, then $N = LF \times 8 = 1.68$.
   Then,

$$N = \frac{\sigma_s + \gamma z - \sigma_T}{c_u} \text{ or } \frac{0 + 20 \times 12.5 - \sigma_T}{25} = 1.68$$

$$\sigma_T = 208 \text{ kPa}$$

b. For a surface surcharge load of 75 kPa while maintaining the same settlement requirements, then,

$$N = \frac{\sigma_s + \gamma z - \sigma_T}{c_u} \text{ or } \frac{75 + 20 \times 12.5 - \sigma_T}{25} = 1.68$$

$$\sigma_T = 283 \text{ kPa}$$

Worked-out example 12.6: For the tunnel considered in worked-out example 12.4, calculate the transversal surface settlement profile if another tunnel having the same dimensions is excavated at the same depth but 10 m away from the centerline of the first tunnel.

Solution: Using the superposition method, the surface transversal surface settlement can be found for tunnel A (worked-out example 12.4 tunnel) and B. For simplicity, only the most conservative values of S_{max} of 21.6 mm and $i = 5$ obtained using Atkinson and Potts (1979) are considered for this problem.

   Note that past studies (e.g. Suwansawat and Einstein, 2007) have reported that the surface settlements induced when the second tunnel is excavated are smaller than those estimated in Table 12.4. Thus, the transversal settlement profile computed is usually conservative (Fig. 12.13).

Table 12.4 Values computed for the transverse surface settlement profile.

Y	Tunnel A	Tunnel B	Both tunnels
−30	3.29E-10	2.74E-16	3.29E-10
−27	1.01E-08	2.78E-14	1.01E-08
−24	2.14E-07	1.97E-12	2.14E-07
−21	3.19E-06	9.71E-11	3.19E-06
−18	3.31E-05	3.35E-09	3.31E-05
−15	2.40E-04	8.05E-08	2.40E-04
−12	1.21E-03	1.35E-06	1.21E-03
−9	4.27E-03	1.58E-05	4.29E-03
−6	1.05E-02	1.29E-04	1.06E-02
−3	1.80E-02	7.35E-04	1.88E-02
0	2.16E-02	2.92E-03	2.45E-02
3	1.80E-02	8.11E-03	2.61E-02
6	1.05E-02	1.57E-02	2.62E-02
9	4.27E-03	2.12E-02	2.54E-02
10	2.92E-03	2.16E-02	2.45E-02
12	1.21E-03	1.99E-02	2.12E-02
15	2.40E-04	1.31E-02	1.33E-02
18	3.31E-05	6.01E-03	6.04E-03
21	3.19E-06	1.92E-03	1.92E-03
24	2.14E-07	4.29E-04	4.29E-04
27	1.01E-08	6.67E-05	6.67E-05
30	3.29E-10	7.25E-06	7.25E-06
33	7.51E-12	5.49E-07	5.49E-07
36	1.20E-13	2.90E-08	2.90E-08

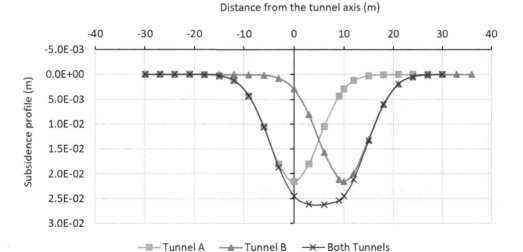

Figure 12.13 Transversal surface settlement profiles of worked-out example 12.5.

Worked-out example 12.7: A tunnel for a new subway line 5 m in diameter is proposed to be excavated at a depth of 15 m in a silty sand deposit (γ_{sat} = 21 kN/m³, E = 25 MPa, k_0 = 0.5, c' = 0 kPa, ϕ = 35°).

a. Derive the Terzaghi (1943) general relationship for computing vertical stresses on a tunnel considering arching theory applied to tunnels in sands.

$$\sigma_v = \frac{\gamma B_1}{k_0 \tan\phi}\left(1 - e^{\left(-k_0 \tan\phi \; D_1/B_1\right)}\right) + \gamma D_2 e^{\left(-k_0 \tan\phi \; D_1/B_1\right)}$$

b. Compute the vertical stress at the tunnel crown considering

   i.   Arching extends to the surface and
   ii. Arching extends only to 5 m above the tunnel crown.

Solution:
a. Terzaghi made the following assumptions based on experimental observations:

   •   Sliding surfaces are assumed to be vertical (sections **ae** and **bf** of Fig. 12.14).
   •   Pressure on the yielding strip is equal to the difference between the weight of the sand located above the strip and at the section (ab) and the shear resistance along the vertical sliding surfaces.

   Figure 12.14 shows a section through the space between two vertical surfaces of sliding. The shearing resistance of the soil is determined as follows:

$$\tau = c' + \sigma \tan\phi'$$

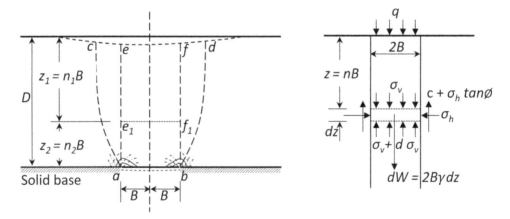

Figure 12.14 Yielding in soil caused by downward movement of a long narrow section (ab) at the base, lines ae and bf are assumed sliding surfaces and (b) free body diagram for a slice of soil in the yielding zone (modified after Terzaghi, 1943).

The unit weight of the soil is γ_{sat}, and the surface of the soil carries a uniform surcharge q per unit of area. The vertical stress on a horizontal section at any depth z below the surface is σ_v, and the corresponding horizontal stress (σ_h) is

$$\sigma_h = \sigma_v \times K_0$$

where K_0 is the earth pressure coefficient.

The weight of the slice with a thickness dz at a depth z below the surface is $2B\,\gamma dz$ per unit of length perpendicular to the plane of the drawing. The slice is acted upon by the forces indicated in Figure 12.14. The sum of the vertical components that act on the slice must be equal to zero for equilibrium as follows:

$$2B\,\gamma dz = 2B\left(\sigma_v + d\sigma_v\right) - 2B\sigma_v + 2c'dz + 2K_0\sigma_v \times dz \times \tan\phi'$$

or

$$\frac{d\sigma_v}{dz} = \gamma - \frac{c'}{B} - K_0\sigma_v\frac{\tan\phi'}{B}$$

Considering $\sigma_v = q$ for $z = 0$ and solving this equation, we can obtain

$$\sigma_v = \frac{B(\gamma - c'/B)}{K\tan\phi'}\left(1 - e^{\left(-K\tan\phi'\,{}^z\!/_B\right)}\right) + qe^{\left(-K\tan\phi'\,{}^z\!/_B\right)}$$

Substituting $B = B_1$, $q = \gamma D_2$ and $z = D_1$ for application to square tunnels in sand ($c' = 0$) (Fig. 12.5) gives:

$$\sigma_v = \frac{B(\gamma - c'/B)}{k_0\tan\phi'}\left(1 - e^{\left(-k_0\tan\phi'\,{}^z\!/_B\right)}\right) + qe^{\left(-k_0\tan\phi'\,{}^z\!/_B\right)}$$

This relationship can be simplified further for different cases, considering $c' = 0$ and $q = 0$ as follows:

$$c' > 0 \text{ and } q = 0 : \sigma_v = \frac{B_1(\gamma - c'/B_1)}{k_0\tan\phi'}\left(1 - e^{\left(-k_0\tan\phi\,{}^{D_1}\!/_{B_1}\right)}\right)$$

$$c' = 0 \text{ and } q > 0 : \sigma_v = \frac{B_1\gamma}{k_0\tan\phi'}\left(1 - e^{\left(-k_0\tan\phi'\,{}^{D_1}\!/_{B_1}\right)}\right) + qe^{\left(-k_0\tan\phi\,{}^{D_1}\!/_{B_1}\right)}$$

$$c' = 0 \text{ and } q = 0 : \sigma_v = \frac{B\gamma}{k_0\tan\phi'}\left(1 - e^{\left(-k_0\tan\phi'\,{}^{D_1}\!/_{B_1}\right)}\right)$$

b. The relationship derived may be adopted to compute the vertical stress at the tunnel crown ($c' = 0$):

$$\sigma_v = \frac{\gamma B_1}{k_0\tan\phi'}\left(1 - e^{\left(-k_0\tan\phi'\,{}^{D_1}\!/_{B_1}\right)}\right) + \gamma D_2 e^{\left(-k_0\tan\phi'\,{}^{D_1}\!/_{B_1}\right)}$$

Making a reasonable assumption of a square geometry for the tunnel, then,

$$2B_1 = 2\left[B_0 + H \tan\left(45 + \frac{\phi}{2}\right)\right]$$

$$B_1 = \left[2.5 + 2.5 \tan\left(45 + \frac{35}{2}\right)\right] = 7.3\,\text{m}$$

i. arching extends to the surface, then $D_2 = 0$ and

$$\sigma_v = \frac{\gamma B_1}{k_0 \tan\phi}\left(1 - e^{\left(-k_0 \tan\phi \; D_1/B_1\right)}\right) \quad \text{then}$$

$$\sigma_v = \frac{21 \times 7.3}{0.5 \tan 35}\left(1 - e^{\left(-0.5 \tan 35 \; 15/7.3\right)}\right) = 224.6\,\text{kPa}$$

ii. if aching extends only to 5 m above the tunnel crown, then $D_1 = 5$ m and $D_2 = 10$ m.

$$\sigma_v = \frac{\gamma B_1}{k_0 \tan\phi'}\left(1 - e^{\left(-k_0 \tan\phi' \; D_1/B_1\right)}\right) + \gamma D_2 e^{\left(-k_0 \tan\phi' \; D_1/B_1\right)} \quad \text{then}$$

$$\sigma_v = \frac{21 \times 7.3}{0.5 \tan 35}\left(1 - e^{\left(-0.5 \tan 35 \; 5/7.3\right)}\right) + 21 \times 10 \times e^{\left(-0.5 \tan 35 \; 5/7.3\right)}$$

$$\sigma_v = 91.95 + 165.27 = 257.22\,\text{kPa}$$

Worked-out example 12.8: For the tunnel considered in worked-out example 12.4 (depth 10 m and diameter of 5 m) excavated in a clayey sand ($\gamma_{sat} = 22$ kN/m³, $K_0 = 0.5$, $\upsilon = 0.3$), calculate the vertical pressure on the tunnel supports considering arching extends to the surface and long-term conditions where $c' = 5$ kPa and $\phi' = 15°$.

Solution: If arching extends to the surface, then $D_2 = 0$ and $c' \neq 0$, then

$$\sigma_v = \frac{B_1(\gamma - c'/B_1)}{K_0 \tan\phi'}\left(1 - e^{\left(-K_0 \tan\phi \; D_1/B_1\right)}\right)$$

Making a reasonable assumption of a square geometry for the tunnel, then,

$$2B_1 = 2\left[B_0 + H \tan\left(45 + \frac{\phi}{2}\right)\right]$$

Alternatively,

$$B_1 = \left[2.5 + 2.5 \tan\left(45 + \frac{15}{2}\right)\right] = 5.76\,\text{m}$$

Then,

$$\sigma_v = \frac{5.76(22-5/5.76)}{0.5\tan 15}\left(1-e^{\left(-0.5\tan 15\ \ ^{10}\!/_{5.76}\right)}\right) = 188.5\,\text{kPa compared to the } \sigma_v = \gamma z = 220\,\text{kPa}$$

Worked-out example 12.9: For an immersed tunnel with the cross-section displayed in what follows, determine the total weight and volume required in ballast to counter the uplift during sinking of the different elements into position (Fig. 12.15).

You may assume unit weight of reinforced concrete = 24.13 kN/m³, ballast = 22.1 kN/m³ and a flotation margin of 5%.

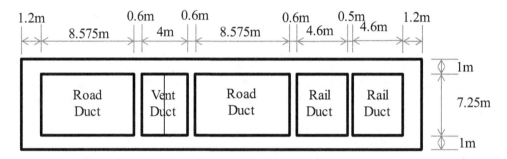

Figure 12.15 Diagram of an immersed tunnel.

Solution: Dimensions of the tunnel: height = 9.25 m and width = 35.05 m.

Then the volume of water displaced = 9.25 × 35.05 ×1 = 324.21 m³ (per meter length of segment).

$$W_{\text{water, displaced}} = \gamma_w \times V_{\text{tunnel, outline}} = 9.81 \times 324.21 = 3180.5\,\text{kN}$$

Simplifying the openings of the tunnel into rectangles, volume of the tunnel is computed as:

Cross-sectional area of the openings = (8.575 + 4 + 8.575 + 4.6 + 4.6)
$$\times\ 7.25 = 220.04\ \text{m}^2$$
$$V_{\text{tunnel}} = (324.21 - 220.04) \times 1 = 104.2\ \text{m}^3$$
$$W_{\text{tunnel}} = 104.2 \times 24.13 = 2514.35\ \text{kN}$$

Then

$$W_{\text{tunnel, concrete}} + W_{\text{ballast}} = W_{\text{water, displaced}} \times f_m$$
$$2514.35 + W_{\text{ballast}} = 3180.5 \times 1.05$$
$$W_{\text{ballast}} = 825.2\ \text{kN}$$
$$W_{\text{ballast}} = \gamma_b \times V_{\text{ballast}}$$
$$V_{\text{ballast}} = 825.2 / 22.1 = 37.3\ \text{m}^3$$

References

Atkinson, J.H. & Potts, D.M. (1979) Subsidence above shallow tunnels in soft ground. *Journal of Geotechnical Engineering, American Society of Civil Engineers*, GT4, 307–325.

Broms, B. &Bennermark, H. (1967) Stability of clays at vertical openings. *Journal of Geotechnical Engineering ASCE*, 93(SM1), 71–74.

Clough, G.W. & Schmidt, B. (1981) Excavations and tunneling. Chapter 8. In: Brand, E.W. & Brenner, R.P. (eds) *Soft Clay Engineering*. Elsevier, Amsterdam.

Dimmock, P.S. & Mair, R.J. (2007) Estimating volume loss for open faced tunnels in London clay. *Proceedings of Institution of Civil Engineers – Geotechnical Engineering*, 160(1), 13–22.

ITA-AITES Working Group 'Research' (2007) ITA/AITES Report 2006 on Settlements induced by tunneling in soft ground. *Tunneling and Underground Space Technology*, 22, 119–149.

Mair, R.J. & Taylor, R.N. (1997) Theme lecture: Bored tunneling in the urban environment. *Proceedings of the 14th International Conference on Soil Mechanics and Foundation Engineering*, Hamburg, 4, CRC Press, pp. 2353–2385.

Mair, R.J., Taylor, R.N. & Burland, J. (1996) Prediction of ground movements and assessment of risk of building damage due to bored tunneling. *Conference on Geotechnical Aspects of Underground Construction in Soft Ground*, London.

Marshall, A.M., Farrell, R., Klar, A. & Mair, R. (2012) Tunnels in sands: The effect of size, depth and volume loss on greenfield displacements. *Geotechnique*, 62(5), 385–399. http://dx.doi.org/10.1680/geot.10.P.047.

O'Reilly, M.P. & New, B.M. (1982) Settlements above tunnels in the UK-their magnitude and prediction. *Tunneling*, 82, 173–181.

Peck, R.B. (1969) Deep excavations and tunneling in soft ground. *Proceedings of the 7th International Conference on Soil Mechanics and Foundation Engineering*, Mexico City, pp. 225–290.

Suwansawat, S.& Einstein, H.H.(2007) Describing settlement troughs over twin tunnels using a superposition technique. *Journal of Geotechnical and Geoenvironmental Engineering*, 133(4), https://doi.org/10.1061/(ASCE)1090-0241(2007)133:4(445).

Terzaghi, K. (1943) *Theoretical Soil Mechanics*. John Wiley & Sons, Inc., ISBN:9780471853053, p. 510.

Vorster, T.E.B., Klar, A., Soga, K. & Mair, R.J. (2005) Estimating the effects of tunneling on existing pipelines. *Journal of Geotechnical Geoenvironmental Engineering*, 131(11), 1399–1410.

Chapter 13

Pavement infrastructure (roads and rails)

This chapter introduces different approaches for designing typical pavements applied to roads and rail infrastructure. Pavements are engineered structures used for the transportation of people and goods in our daily lives. The important function of the pavements is to withstand and distribute the loads from the traffic (vehicles or trains) to underlying subgrade soil. Pavement design is a process of selecting appropriate materials for pavement and surface to ensure that it performs satisfactorily and requires only minimum maintenance under the expected vehicle loads for the required design life. The design of pavements depends on many factors such as vehicle wheel load, configuration of vehicle wheel, volume of traffic and subgrade soil strength.

13.1 Design of pavements

The pavements can be classified as either flexible or rigid based on their structural performance. Flexible pavements consist of granular pavement materials with bituminous surfacing, and rigid pavements are generally concrete pavements (see Fig.13.1).

The design of flexible pavements is generally based on California Bearing Ratio (CBR) tests results. The CBR value is determined by penetration tests and is defined as the rate of the force per unit area required to penetrate a soil mass with a standard circular solid steel plunger of 50 mm diameter at rate of 1 mm/min to that required for the penetration of standard material. The standard material is usually confined crushed rock (i.e. CBR of 100%). The standards load for 2.5 mm and 5 mm penetration are generally 13.2 kN and 20.0 kN, respectively.

The stiffness or elastic modulus of the subgrade can be determined from CBR value using the empirical relationship proposed by Powell et al. (1984):

$$E = 17.6(\text{CBR \%})^{0.64} \text{ MPa} \tag{13.1}$$

Flexible pavements may be designed using laboratory-soaked CBR, field CBR or CBR from undisturbed samples. However, laboratory-soaked CBR is generally used for designing the thickness of the flexible pavements. Figure 13.2 shows the thickness of material required over the in-situ subgrade as function of Equivalent Standard Axles (ESA).

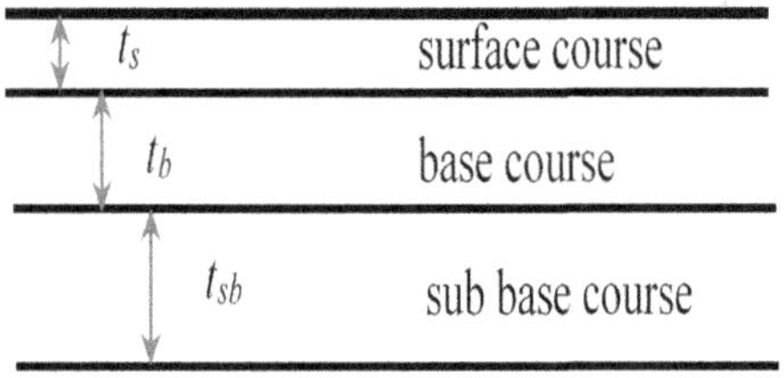

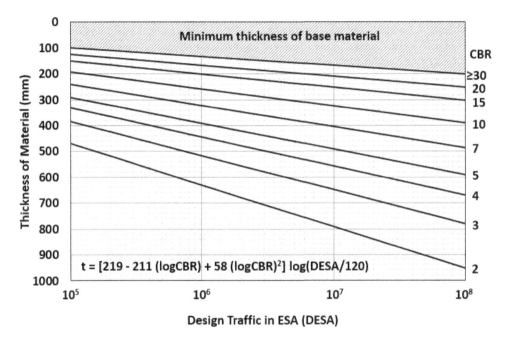

Figure 13.1 Cross section of road pavement.

Figure 13.2 Design chart for granular pavement with thin bituminous surfacing (modified after Austroads, 2017).

13.2 Rail track modulus and settlement

The current worldwide trend toward increased axle loads and faster trains has resulted in an increased damage to heavy-haul tracks. The ability to accurately assess the structural condition of the track has become very important. One important parameter for characterizing the condition of the track substructure is the track modulus.

A number of theoretical models have been proposed for the calculation of track modulus based on load versus deflection relationships, yet there is no consensus on the best or most accurate method. The most commonly known method assumes the rail track assembly to be a beam on an elastic foundation (i.e. the Winkler model), as illustrated in Figure 13.3. A vertical force (P) applied by a wheel produces a vertical rail deflection (w). Therefore, the track stiffness (k), taken at a point as the wheel passes directly overhead, is defined as follows:

$$k = \frac{P}{w} \tag{13.2}$$

where k = track stiffness; P = vertical force applied by a wheel; and w = measured deflection of the rail.

From this, the track modulus (u) can be determined as:

$$u = \frac{k^{4/3}}{(64EI)^{4/3}} \tag{13.3}$$

where E is the Young's modulus of the rail and I is the rail's moment of inertia.

Another approach to determine the overall track modulus employs the use of the equivalent method in which the modulus of granular layer ($\bar{E}$) of the whole track (Fig. 13.4) substructure can be estimated as follows:

$$\bar{E} = \frac{H_b + H_c + H_f}{\left(\dfrac{H_b}{E_b} + \dfrac{H_c}{E_c} + \dfrac{H_f}{E_f} \right)} \tag{13.4}$$

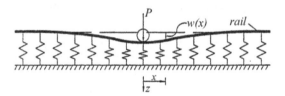

Figure 13.3 Typical rail-in-track subjected to a wheel load.

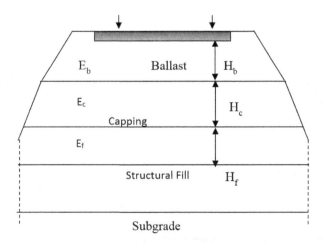

Figure 13.4 Schematic diagram of granular layers representing the elastic modulus of each layer.

where E_b, E_c and E_f are the elastic modulus of the ballast, capping and structural fill, respectively. H_b, H_c and H_f are the thicknesses of the ballast, capping, and structural fill, respectively.

Worked-out example 13.1: The following results are obtained from a standard CBR test for the subgrade soil. Determine the CBR value of the subgrade soil.

Penetration (mm)	0.6	1.2	1.8	2.4	3.0	3.6	4.2	4.8	5.4	6
Load kN		3.9	6.5	8.5	10.3	11.2	11.8	12.6	13.1	13.8	14.3

The standard load for 2.5 mm and 5 mm penetration is 13.2 kN and 20.0 kN, respectively.

Solution: The standard penetration curve is shown in Figure 13.5. From Figure 13.5, the load corresponding to 2.5 mm and 5 mm penetration is 10.5 kN and 13.4 mm, respectively.

The CBR value for 2.5 mm penetration = 10.5/13.2 = 79.5%.
The CBR value of 5 mm penetration = 13.4/20.0 = 67%.
The higher value is considered as CBR for the subgrade = 79.5%.

Worked-out example 13.2: A new transport link is proposed. The traffic forecast indicated that a new road is expected to operate for a traffic of 6×10^6 ESA. During the site investigation campaign, samples were collected from trial pits, and a number of CBR tests were conducted. The CBR test results indicate the subgrade has a CBR of 10%.

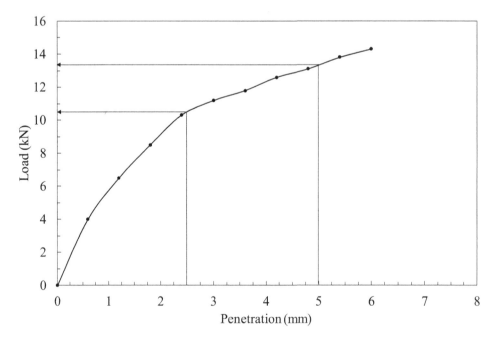

Figure 13.5 CBR results: load versus penetration.

Determine the minimum thickness of the base course and subbase course layers of the new road pavement. Assume a CBR>30% for base course.

Solution:

Design CBR = 10%

Design traffic: 6×10^6 ESA

From Figure 13.1, the total thickness of the material over subgrade is 300 mm. It is suggested to have a CBR>30% for base course materials. From Figure 13.2, corresponding to a CBR of 30%, the thickness of the base course is 175 mm.

Therefore, the total thickness of the subbase course is 300–175 = 125 mm.

Worked-out example 13.3: Determine overall track modulus for a given track structure and settlement with the following information:

Ballast layer: thickness, H_b = 350 mm; modulus of ballast: E_b= 210 MPa
Capping: thickness, H_c = 150 mm; modulus of capping: E_b= 110 MPa
Structure fill: thickness, H_f = 700 mm; modulus of structural fill: E_b= 40 MPa

Solution:

Calculation procedure:

Equivalent modulus of granular layer ($\bar{E}$) is calculated as:

$$\bar{E} = \frac{H_b + H_c + H_f}{\left(\dfrac{H_b}{E_b} + \dfrac{H_c}{E_c} + \dfrac{H_f}{E_f} \right)}$$

where E_b, E_c and E_f are the elastic modulus of the ballast, capping and structural fill. H_b, H_c and H_f are the thicknesses of ballast, capping and structural fill, respectively.

$$\bar{E} = \frac{0.35 + 0.15 + 0.7}{\dfrac{0.35}{210} + \dfrac{0.15}{110} + \dfrac{0.7}{40}} = 58.45 \text{ MPa}$$

Assume the equivalent dynamic stress at the sleeper/ballast is 550 kPa (Indraratna and Ngo 2018). The average strain (ε_{ave}) of the equivalent granular medium is calculated as the stress at sleeper/ballast interface divided by the equivalent modulus ($\bar{E}$), determined as:

$$\varepsilon_{ave} = \frac{\sigma_{dyn}}{\bar{E}} = \frac{550}{58.45 \times 10^3} \times 100 = 0.941\%$$

The elastic settlement of granular medium is then predicted using the total height (H = 1.2 m) and determined as:

$$S_{total} = \varepsilon_{ave} \times H = 0.941/100 \times 1.2 = 0.01129 \text{ (m)} = 11.29 \text{ mm}$$

Worked-out example 13.4: A new track extension is proposed to be built on a homogeneous and uniform clay subgrade. The track is subjected to regular heavy wheel loads whose details are summarized in Table 13.1. Determine a thickness of granular layer for track foundation. (Note: this example is similar to the one presented by Li and Selig, 1998a.)

Table 13.1 Input parameters for determining the thickness of the granular layer based on the performance of the subgrade (Li and Selig, 1998b).

Design parameters	Values
Design criteria	Allowable subgrade plastic strain for the design period, ε_{pa} = 2%
	Allowable settlement of subgrade in design period, ρ_a = 25 mm
	Minimum granular layer height = 0.45 m
	Impact factor method = AREA (1974)
	Subgrade capacity method = Li and Selig (1998b)
Rail and sleeper properties	Not needed
Traffic conditions	Static wheel load, P_s = 173 kN
	Velocity, V = 64 km/h
	Design tonnage, T = 64 MGT
	Wheel diameter, D = 0.97 m
Granular material characteristics	Resilient modulus, $E_b = E_c = E_s$ = 276 MPa
Subgrade soil characteristics	Soil type, CH (fat clay)
	Soil compressive strength, σ_s = 90 kPa
	Subgrade modulus, E_s = 14 MPa
	Thickness = 1.5 m

Solution:

Step 1: Dynamic amplification factor (*IF*) calculation based on AREA (1974) method.

The American Railroad Engineering Association (AREA, 1974) introduced a simple mathematical expression to calculate the value of *IF* based on the results of in-situ measurements of dynamic wheel loads from train cars with known static wheel loads. All the measurements were undertaken on standard-gauge tracks (1435 mm) in the U.S. The value of *IF* is set to be a function of vehicle speed V (km/h) and wheel diameter D (mm).

$$IF = 1 + 5.21 \frac{V}{D}$$

$$IF = 1 + 5.21 \frac{V}{D} = 1 + 5.21 \frac{64}{970} = 1.34$$

Step 2: Dynamic wheel load (P_d) calculation:

$$P_d = 1.34 \times 173 = 231.82 \text{ kN}$$

Step 3: Number of load cycle (N) calculation:

$$N = \frac{T}{8P_s} = \frac{\left(60 \times 10^6\right) \times 9.81}{8 \times 173} = 425{,}289$$

Step 4: Select the values of a, m and b for CH soil from Table 13.2:

$a = 1.2$; $m = 2.4$; $b = 0.18$

Table 13.2 Values of soil parameters a, b and m for calculating the plastic shear strain of subgrade for four types of soil (Li and Selig, 1998a).

Soil type	a	b	m
CH (fat clay)	1.20	0.18	2.40
CL (lean clay)	1.10	0.16	2.00
MH (elastic silt)	0.84	0.13	2.00
ML (silt)	0.64	0.10	1.70

Step 5: Calculation for the first design procedure (preventing local shear failure of subgrade):

Step 5.1: Calculate allowable deviator stress on subgrade (σ_{da}) using

$$\sigma_{da} = \sigma_s \left[\left(\frac{\varepsilon_{pa}}{aN^b} \right)^{(1/m)} \right]$$

$\varepsilon_p = 2\%$ and $\sigma_s = 90$ kPa is used from the input parameters.

$$\sigma_{da} = \sigma_s \left[\left(\frac{\varepsilon_p}{aN^b} \right)^{(1/m)} \right] = 90 \left[\left(\frac{2}{1.2 \times 425289^{0.18}} \right)^{(1/2.4)} \right] = 42.1 \text{ kPa}$$

Step 5.2: Calculate the strain influence factor (I_ε) from $I_\varepsilon = \frac{\sigma_d A}{P_d}$

$$I_\varepsilon = \frac{42.1 \times 0.645}{232.47} = 0.117$$

Step 5.3: Determine the (H/L) from Figure 13.6 corresponding to the granular material modulus $E_s = 276$ MPa and subgrade modulus 14 MPa:

H/L = 5.07
H = 5.07 × 0.152 = 0.771 m

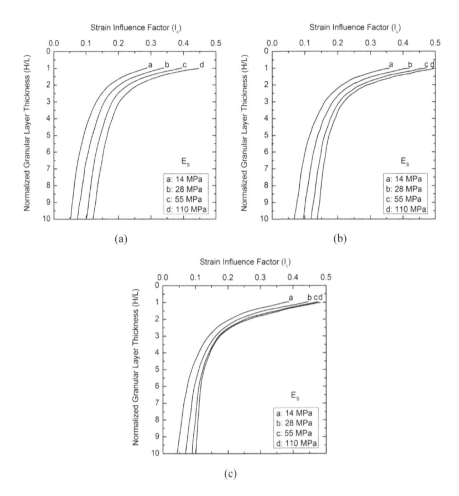

Figure 13.6 Granular layer thickness design chart for preventing progressive shear failure: (a) E_b = 550 MPa, (b) E_b = 280 MPa and (c) E_b = 140 MPa (modified after Li and Selig 1998a).

References

AREA. (1974) *American Railway Engineering Association: Manual for Recommended Practice*. AREMA, Washington, DC.

Austroads. (2017) *Guide to Pavement Technology Part 2: Pavement Structural Design*. Austroads Publication No. AGPT02-17, 283.

Indraratna, B. & Ngo, T. (2018) *Ballast Railroad Design: Smart-Uow Approach*. CRC Press, Taylor & Francis Group, Netherlands.

Li, D. & Selig, E.T. (1998a) Method for railroad track foundation design. I: Development. *Journal of Geotechnical and Geoenvironmental Engineering*, 124(4), 316.

Li, D. & Selig, E.T. (1998b) Method for railroad track foundation design. II: Applications. *Journal of Geotechnical and Geoenvironmental Engineering*, 124(4), 323.

Powell, W.D., Potter, J.F., Mayhew, H.C. & Nunn, M.E. (1984) *The Structural Design of Bituminous Roads*. Department of Environment, Department of Transport, TRRL Report LR1132, Crowthorne, Berks.

Chapter 14

Unsaturated soil mechanics and applications

This chapter introduces a number of different problems related to applied topics in unsaturated soil mechanics.

14.1 Quantities describing water content

The amount of water in soil can be expressed in terms of mass, known as the gravimetric water content, or in terms of volume, known as the volumetric water content. The gravimetric water content, typically referred to simply as water content, is expressed by the mass ratio of water and solid particles in any unit soil volume (see Chapter 2):

$$w = \frac{m_w}{m_s} \tag{14.1}$$

whereas the volumetric water content (θ_w) is defined by the volume ratio of water and total volume of soil:

$$\theta_w = \frac{V_w}{V} \tag{14.2}$$

In unsaturated soil, the amount of water in the soil is often represented by the degree of saturation (S_r), expressed by the ratio of the volume of water and volume of the voids (see Chapter 2):

$$S_r = \frac{V_w}{V_v} \tag{14.3}$$

The different quantities can be interrelated by:

$$\theta_w = nS_r = \frac{wG_s}{1+e} \tag{14.4}$$

where n represents porosity, G_s is the specific gravity of solids, and e is the void ratio. Note that the amount of water stored in the voids can also be represented by the water ratio (e_w), and it is defined as the ratio of the volume of water and volume of the solids, or

$$e_w = \frac{V_w}{V_s} \tag{14.5}$$

14.2 Soil water retention behavior

14.2.1 Suction potential

Suction potential is the potential energy of water relative to pure water in a reference state. It quantifies the tendency of water to move from one area to another due to osmosis, gravity, mechanical pressure or capillary action. From a thermodynamic standpoint, total suction (s) can be quantitatively described by Kelvin's equation, given in Eq. 14.6:

$$s = -\frac{RT}{v_{v0}\omega_v}\ln\left(\frac{u_v}{u_{v0}}\right) \tag{14.6}$$

where s is the total suction (kPa), R is the universal gas constant (J/(mol·K)), T is the absolute temperature (K), v_{w0} is the specific volume of water (m^3/kg), ω_v is the molecular mass of water vapor (g/mol), u_v is the partial pore-water vapor (kPa) and u_{v0} is the saturated pressure of water vapor over a flat surface of pure water at the same temperature (kPa). The term (u_v/u_{v0}) is called relative humidity, or RH (%). Two components can influence the total suction: they are the pressure or capillary potential (u_a-u_w) and the osmotic potential (π), as shown in Eq. 14.7:

$$s = \left(u_a - u_w\right) + \pi \tag{14.7}$$

14.2.1.1 Capillary potential or matric suction

While the water pressure in saturated soil (i.e. below the water table) is positive and the liquid phase is in compression, in unsaturated soil (i.e. above the water table), the hydrostatic water pressure is negative, and therefore the liquid phase is in tension. This tension, designated as matric suction, is a result of the interaction of three phases (i.e. soil grains, water and air), and it represents the difference in pressure between the soil water and the air, as in Eq. 14.8:

$$s = \left(u_a - u_w\right) \tag{14.8}$$

Where u_a and u_w are the air and water pressure, respectively.

The matric suction represents the combined effect of adsorption between the soil and water molecules and capillary action within the soil matrix. While the adsorbed water is tightly bonded to the soil particles and could be considered part of a solid skeleton, capillary action is a result of liquid surface tension. The difference between water and air pressures is related to the curvature of the interface of air and water phases (radii r_1 and r_2) and the surface tension T_s (Fig. 14.1), or

$$s = \left(u_a - u_w\right) = T_s\left(\frac{1}{r_1} + \frac{1}{r_2}\right) \tag{14.9}$$

The force derived by the capillary action acting on the grain contacts can easily be quantified using a simple two-spheres model (Fig. 14.1).

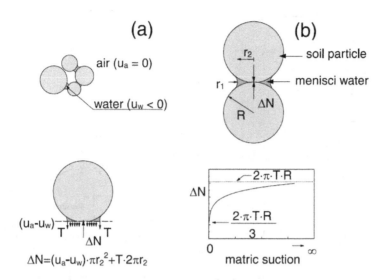

Figure 14.1 Schematic illustration of micro-scale models for unsaturated spherical particles (Mancuso *et al.*, 2002, reproduced with permission from *Canadian Geotechnical Journal*).

14.2.1.2 Osmotic potential or osmotic suction

The osmotic potential of water is linked to the dissolved solutes (i.e. salts) in soil water. Pure water is usually defined as having a solute potential of zero that increases in a negative order for increasingly higher concentrations of solute in the pore fluid. Because osmotic suction can only be observed if the soil water comes in contact with pure water through a perfect semipermeable membrane that should only allow water to pass through and not the solutes, it is not usually measured in soils.

The matric suction is widely accepted to be the governing component of suction in describing the mechanical behavior of unsaturated soils, and it has been used to formulate the mechanical constitutive models for these materials. In the following sections, the term "suction" refers to matric suction unless stated otherwise.

14.2.2 Soil water retention curve

To adequately describe the hydraulic properties and understand the volumetric behavior of unsaturated soil, it is necessary to establish a relationship between suction and the amount of water in the soil, which can be expressed in terms of either water content (mass), volumetric water content or degree of saturation (volume).

This relationship is often referred to as the soil-water retention curve or SWRC. The desaturation process expressed by the SWRC can be divided into three main ranges, that is, the boundary effect zone, the transition zone and the residual zone (Fig. 14.2a). The first transition point represents the air-entry value (AEV) whereas the second represents the onset of residual zone.

The shape and location of transition points is strongly dependent on the type of soil, particle size distribution and its structure. For instance, the SWRC curves of soils that are composed predominantly of sand show smaller AEV values (in the order of 10 kPa), sharper

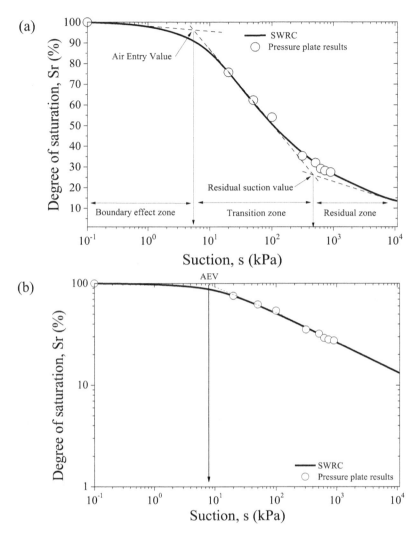

Figure 14.2 SWRC of a silty sand using (a) semi-logarithm scale and (b) double logarithm scale (modified after Heitor, 2013).

slope curves with limited transition zones, whereas soils composed of clay show larger AEV and smoother curves with larger transition ranges.

14.2.2.1 Influence of bimodal porosity

Compacted soils, particularly those compacted at the dry side of optimum moisture content, often exhibit different pore size sets. This is usually referred to as double or bimodal porosity. The SWRC of soils that show bimodal porosity features are notorious because they show two different air entry values, associated to the macropore and micropore ranges (Fig. 14.3).

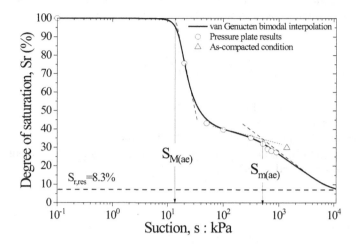

Figure 14.3 Bimodal SWRC of a silty sand (modified after Heitor, 2013).

While in the low suction range, the hydraulic behavior is mainly controlled by the macroporosity (inter-aggregate porosity) for suction values exceeding the micropore air entry ($S_{m(ae)}$) the microporosity (intra-aggregate porosity) governs.

14.2.2.2 Models

There have been different empirical models proposed to model the behavior of SWRC. The most commonly use are those proposed by Fredlund and Xing (1994), represented by Eq. 14.10, and van Genutchen (1980,) represented by Eqs. 14.11 and 14.12.

$$\theta_v = C(\psi)\frac{\theta_s}{\left\{\ln\left[e+(\alpha s)^n\right]\right\}^m} \quad \text{with } C(\psi) = 1 - \frac{\ln\left(1+\dfrac{s}{\psi_r}\right)}{\ln\left(1+\dfrac{10^6}{\psi_r}\right)} \tag{14.10}$$

$$S_{re} = \frac{1}{\left[1+(\alpha s)^n\right]^m} \quad \text{with } m = 1 - \frac{1}{n} \tag{14.11}$$

$$S_{re} = \frac{\theta-\theta_r}{\theta_s-\theta_r} = \frac{nS_r-\theta_r}{\theta_s-\theta_r} \tag{14.12}$$

where S_{re} represents the effective degree of saturation (macropores) that can be expressed in terms of porosity (n), residual volumetric water content (θ_r) and saturation volumetric water content (θ_s); α, n and m are fitting parameters; ψ_r corresponds to suction value at residual water content; and s is the suction.

To describe the shape of a bimodal SWRC, the unimodal models may be extended to include the macroporosity (fitting parameters α_M, n_M and m_M) and microporosity (fitting parameters α_m, n_m and m_m) ranges as represented in Eq. 14.11. For instance, the van Genutchen (1980) model was extended as follows (Zhang and Chen, 2005):

$$S_{re} = p_M \frac{1}{\left[1+\left(\alpha_M s\right)^{n_M}\right]^{m_M}} + p_m \frac{1}{\left[1+\left(\alpha_m s\right)^{n_m}\right]^{m_m}} \tag{14.13}$$

where parameters p_M and p_m represent the relative percentage of the components with the large-pore series and the small-pore series in the soil mass.

Gallipoli *et al.* (2003) proposed another extension of the van Genutchen (1980) model and included the void ratio and two more empirical parameters (ϕ, ψ) in the function α ($\alpha = \phi e^{\psi}$) and Eq. 14.11 was modified as follows:

$$S_{re} = \frac{1}{\left[1+\left(\phi e^{\psi} s\right)^{n}\right]^{m}} \tag{14.14}$$

Where e represents the void ratio and ϕ, ψ, m and n are material parameters obtained using best-fit methods, such as the least-squares method.

14.3 Shear strength

Changes in suction can substantially influence the shear strength of soil. There is a general agreement that an increase of suction under post compaction conditions results in an increase in the shear strength. Furthermore, Vanapalli *et al.* (1996) suggested that the magnitude of this increase in shear strength is associated with different ranges of the soil water retention curve. By considering the independent stress variables approach, the shear strength equation for unsaturated soil can be formulated as a linear combination of the two independent stress state variables (Fredlund et al., 2012) shown in Eq.14.15:

$$\tau = c' + \left(\sigma_f - u_a\right)\tan\phi' + s\tan\phi^b \tag{14.15}$$

where c' and ϕ' are the cohesion and friction angle for the soil in a saturated state and ϕ^b is the friction angle that reflects the increase in strength caused by an increase in suction.

Subsequent studies demonstrated that ϕ^b was not constant, as initially anticipated. Indeed, ϕ^b was found to be equivalent to ϕ' before AEV, and beyond that it decreased with increasing suction. A new equation was proposed by Vanapalli *et al.* (1996) and then Fredlund *et al.* (1996) to incorporate this change in the term $\tan\phi^b$ in Eq. 14.15 as follows:

$$\tan\phi^b = \frac{\theta - \theta_r}{\theta_s - \theta_r}\tan\phi' \tag{14.16}$$

where θ_s, θ_r and θ_w are the saturated, residual and current volumetric water content, respectively.

Eq.14.15 implicitly assumes that the friction angle ϕ' (commonly taken as the saturated friction angle) is independent of suction. Past experimental data reported by Delage and Graham (1995) demonstrated that this assumption was not entirely correct, i.e. cohesion and friction angle increased with suction. On the other hand, considering Bishop's effective stress concept for unsaturated soil (Bishop, 1959), the shear stress criterion expressed in Eq. 14.15 can be rewritten as follows:

$$\tau = c' + \left[\left(\sigma_f - u_a \right) + \frac{\tan \phi^b}{\tan \phi'} s \right] \tan \phi' \qquad (14.17)$$

where the original Bishop's parameter χ is now given by $\tan \phi^b / \tan \phi'$.

Worked-out example 14.1: An undisturbed sample of silty sand was collected on site, and the soil water retention behavior was determined in the laboratory using a pressure plate apparatus and filter paper tests. The results obtained are shown in Table 14.1.

Draw the soil water retention curve and determine the air entry and residual values.

Solution: Figure 14.4 shows the values obtained for the SWRC plotted in a semi-logscale. The air entry and residual values can be read from Figure 14.4 and are as follows:

Air entry value = 0.55 kPa
Residual suction value = 420 kPa

Table 14.1 SWRC results obtained for the clay.

Water content (%)	Matric suction (kPa)
4.9	1702
5.3	1019
6.1	522
8.0	141
10.9	32.2
13.4	10.2
15.8	3.2
18.0	1.1
19.5	0.3
19.5	0.1

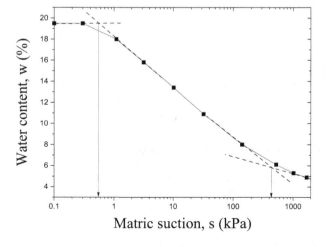

Figure 14.4 SWRC for the sandy silt undisturbed specimen semi-log scale.

Worked-out example 14.2: A new dam is proposed to be built using a local available material, i.e. silty sand. A sample of the soil is compacted to suit the required end-product specifications, and the water retention behavior is determined subsequently using pressure plate tests. The results obtained are shown in Table 14.2. Plot the SWRC and determine the empirical parameters to fit a suitable model.

Table 14.2 SWRC results obtained for the clay.

Suction (kPa)	Degree of saturation (%)
0.1	100
20	75.70796
50	43.25604
100	39.65467
310	35.30136
510	32.01659
620	29.12758
730	28.13819
900	27.46541

Solution: The results can be plotted as follows (Fig. 14.5):

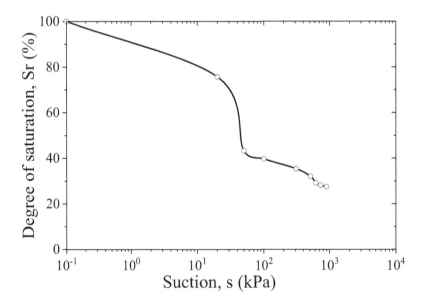

Figure 14.5 SWRC for the sandy silt compacted specimen in semilog scale.

The results seem to indicate that a bimodal model may be more suitable to fit the data. Consider the model proposed by Zhang and Chen (2005):

$$S_{re} = p_M \frac{1}{\left[1+\left(\alpha_M s\right)^{n_M}\right]^{m_M}} + p_m \frac{1}{\left[1+\left(\alpha_m s\right)^{n_m}\right]^{m_m}}$$

Given that the degree of uncertainty is similar for all the unknown parameters, the unweighted least-square method is appropriate. The experimental values may be listed in a Microsoft Excel spreadsheet (Table 14.3), and the vertical deviation of the ith point from the smooth curve is

$$\text{vertical deviation} = y_{i-\text{observed}} - y_{i-\text{calculated}}$$

According to the least-square criterion, the square sum of the n-points deviations should be minimal. For every i^{th} point, the square difference between the measured y_i and the calculated y_{i-m} values can be computed (Table 14.3) by:

$$\left(y_i - y_{i-m}\right)^2 = \sum_{i=1}^{n}\left[y_i - \left(p_M \frac{1}{\left[1+\left(\alpha_M s\right)^{n_M}\right]^{m_M}} + p_m \frac{1}{\left[1+\left(\alpha_m s\right)^{n_m}\right]^{m_m}}\right)\right]^2$$

To perform the least-square method in a Microsoft Excel spreadsheet, we need to choose "data analysis" in the Data tab and then select the option for Solver (Fig. 14.6). The target cell

Table 14.3 Summary of the measured and calculated values.

X	y_i	y_{i-m}	$(y_i - y_{i-m})^2$
0.1	100.000	99.840	0.026
20	75.708	66.576	83.394
50	43.256	53.563	106.234
100	39.655	44.998	28.553
310	35.301	33.645	2.743
510	32.017	29.576	5.956
620	29.128	28.116	1.024
730	28.138	26.950	1.413
900	27.465	25.525	3.767

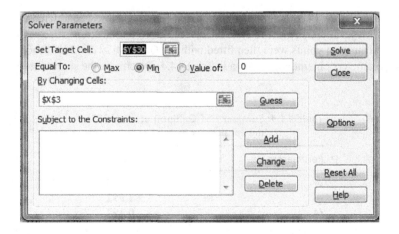

Figure 14.6 Screenshot of the Microsoft Excel solver plugin.

represents the sum of the square differences, and the changing cells represent the parameters α, m and n. Once the appropriate cells in the spreadsheet are selected by clicking "solve," the solver tool looks up in a matter of seconds the best combination of the selected parameters and updates their values in the spreadsheet.

The fitted model with parameters determined can then be plotted together with the SWRC data, as in Figure 14.7.

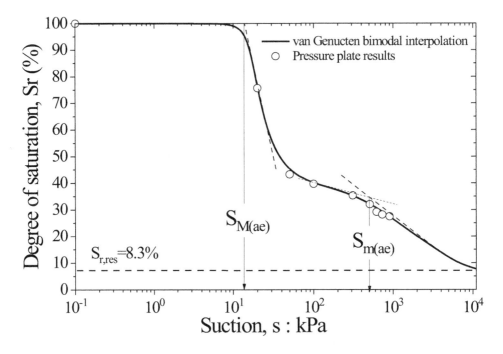

Figure 14.7 SWRC for the sandy silt compacted specimen and bimodal model fit.

Worked-out example 14.3: A series of static compaction tests was conducted using a clay soil. For each specimen, degree of saturation, void ratio and suction were recorded. The different experimental points were then fitted with the Gallipoli *et al.* (2003) model, and the empirical parameters obtained are shown in Table 14.4. Determine the suction level obtained

Table 14.4 Summary of Gallipoli et al. (2003) model parameters

Parameters	Value
ϕ	0.004736
ψ	2.992
M	0.197
N	1.64

for a specimen prepared at a water content of 0.27 and compacted with two vertical stress levels to achieve a void ratio of 1.6 and 1.2. Assume specific gravity of 2.7.

$$S_{re} = \frac{1}{\left[1+\left(\phi e^{\psi} s\right)^{n}\right]^{m}}$$

Solution: The problem describes a typical mechanical wetting path, for which there is a reduction in void ratio for larger static stress. As compaction takes place under constant water content, the degree of saturation can be determined using the basic weight–volume relationship, as follows:

$eS_r = wG_s$ or

$S_r = \dfrac{wG_s}{e}$, hence

$S_{r(e=1.6)} = \dfrac{0.27 \times 2.7}{1.6} = 0.456$

$S_{r(e=1.2)} = \dfrac{0.27 \times 2.7}{1.2} = 0.608$

Using the Gallipoli *et al.* (2003) model, then graphically (Fig. 14.8):

$s_{(e=1.6)} = 570 \text{ kPa}$
$s_{(e=1.2)} = 550 \text{ kPa}$

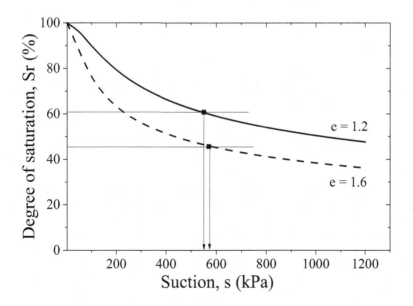

Figure 14.8 SWRC for the compacted specimens for void ratios of 1.2 and 1.6.

Worked-out example 14.4: A series of constant suction triaxial tests was conducted for a clay at a net stress of 100 kPa using an apparatus adopting axis translation technique. Determine the shear strength of the soil at a suction level of 250 kPa, knowing that the volumetric water content at failure was 0.29 and the saturated shear strength parameters, c' and ϕ', were 25 kPa and 10°, respectively.

The soil water retention behavior was determined separately using pressure a plate apparatus and a filter paper method, and the volumetric water content corresponding to saturated (θ_s) and residual state (θ_r) were 0.35 and 0.24, respectively.

Solution: As the water retention behavior data is available, then Vanapalli *et al.* (1996) can be used as follows:

$$\tau = c' + \left(\sigma_f - u_a\right)\tan\phi' + s\tan\phi^b$$

where

$$\tan\phi^b = \frac{\theta - \theta_r}{\theta_s - \theta_r}\tan\phi'$$

$$\tan\phi^b = \frac{0.29 - 0.24}{0.35 - 0.24}\tan 10 = 0.08$$

Thus, for a net stress of 100 kPa and suction level of 250 kPa,

$$\tau = c' + (\sigma_f - u_a)\tan\phi' + s\tan\phi^b$$

$$\tau = 25 + (100)\tan 10 + (250 \times 0.08) = 62.7 \text{ kPa}$$

References

Bishop, A.W. (1959)The principle of effective stress. *Teknisk Ukeblad*, 106, 859–863.

Delage, P. & and Graham, J. (1995) Mechanical behaviour of unsaturated soils: Understanding the behaviour of unsaturated soils requires reliable conceptual models, *Proceedings of the First International Conference on Unsaturated Soils (unsat '95)* Paris/France/6–8 September, Volume 3.

Fredlund, D.G. & Xing, A. (1994) Equations for soil water characteristic curve. *Canadian Geotechnical Journal*, 31(3), 521–532.

Fredlund, D.G., Rahardjo, H. & Fredlund, M.D. (2012) *Unsaturated Soil Mechanics in Engineering Practice*. John Wiley & Sons, Hoboken, NJ.

Fredlund, D.G., Xing, A., Fredlund, M.D. & Barbour, S.L. (1996) The relationship of the unsaturated soil shear strength to the soil-water characteristic curve. *Canadian Geotechnical Journal*, 33(3), 440–448.

Gallipoli, D., Gens, A., Sharma, R. & Vaunat, J. (2003) An elasto-plastic model for unsaturated soil incorporating the effects of suction and degree of saturation on mechanical behaviour. *Géotechnique*, 53(1), 123–135.

Heitor, A. (2013) *Assessment of Post-Compaction Characteristics of an Unsaturated Silty Sand*. Doctor of Philosophy Thesis, School of Civil, Mining and Environmental Engineering, University of Wollongong. Available from: https://ro.uow.edu.au/theses/3732.

Mancuso, C., Vassallo, R. &D'Onofrio, A. (2002) Small strain behaviour of a silty sand in controlled suction resonant column-torsional shear tests. *Canadian Geotechnical Journal*, 39, 22–32.

van Genutchen, M.T. (1980) A close form equation for predicting the water permeability of unsaturated soils. *Journal of Soil Science*, 44, 892–898.

Vanapalli, S.K., Fredlund, D.G. & Pufahl, D.E. (1996) The relationship between the soil-water characteristic curve and the unsaturated shear strength of a compacted glacial till. *Geotechnical Testing Journal*, 19(3), 259–268.

Zhang, L.M. & Chen, Q. (2005) Predicting bimodal soil-water characteristic curves. *Journal of Geotechnical and Geoenvironmental Engineering*, 131(5), 666–670.

Chapter 15

Dewatering, radial drainage and vacuum consolidation

This chapter focuses on aspects related to reclamation and marine works.

15.1 Dewatering

For application in reclamation and marine works, typically a number of dewatering methods can be used to control the groundwater level. The selection of the most appropriate method typically depends on the expected drawdown level and the permeability of the ground (Fig. 15.1).

For an optimal design of dewatering systems, three main aspects must be considered, i.e. area to be dewatered or equivalent circle, radius of influence of each well and capacity of the pumps.

The radius of influence of each well can be determined using Sichardt's formula (1928) as follows:

$$R_0 = C \cdot H \cdot \sqrt{k} \qquad (15.1)$$

where R_0 is the distance of influence (m), H is the expected drawdown (m), k is ground permeability (m/s) and C is an empirical factor, generally taken as 1500–2000 $(m/s)^{-1/2}$ for plane flow and 3000 $(m/s)^{-1/2}$ for wellpoints.

In addition, the spacing and location must be carefully investigated to ensure the water level remains at the expected depth. To compute the rate of water flow arriving at each well, the formulation derived in Chapter 4 for well pumping can be used. For instance, Eq. 15.2 represents an unconfined aquifer, whereas Eq. 15.3 represents a confined aquifer ground conditions (see Chapter 4).

$$q = \frac{k\pi \left(H_2^2 - H_1^2 \right)}{ln\left(\dfrac{R_2}{R_1} \right)} \qquad (15.2)$$

$$q = \frac{k2\pi H(H_2 - H_1)}{ln\left(\dfrac{R_2}{R_1} \right)} \qquad (15.3)$$

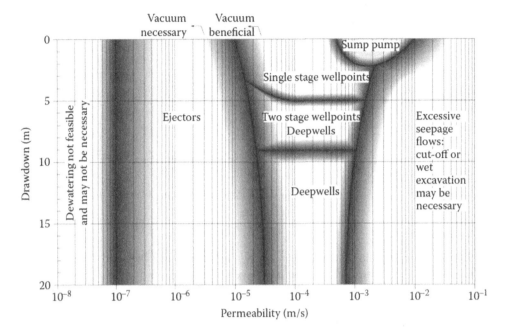

Figure 15.1 Range of application of dewatering techniques (after Powrie, 2014, reproduced with permission from CRC Press).

15.2 Consolidation using vertical drains

Vertical drains can be used to accelerate consolidation of low-lying and reclaimed land, e.g. artificial islands for airports and extension of existing port infrastructure. By creating artificial drainage paths which contribute to a reduction of the length of the drainage path, the drains accelerate the dissipation of the pore water pressure and hence consolidation. There are four types of vertical drains: sand drains, sand compaction/gravel piles, fabric encased sand drains and prefabricated vertical drains (PVDs).

15.2.1 Radial consolidation theory

The three-dimensional consolidation of radial drainage in ideal drains is given by:

$$\frac{\partial \bar{u}}{\partial t} = C_v \left[\frac{\partial^2 u}{\partial z^2} \right] + C_h \left[\frac{\partial^2 u}{\partial r^2} + \frac{1}{r} \frac{\partial u}{\partial r} \right] \tag{15.4}$$

where
u excess pore water pressure at radius r and at the depth z
t time elapsed after the loading applied
C_v coefficient of vertical consolidation
C_h coefficient of horizontal consolidation

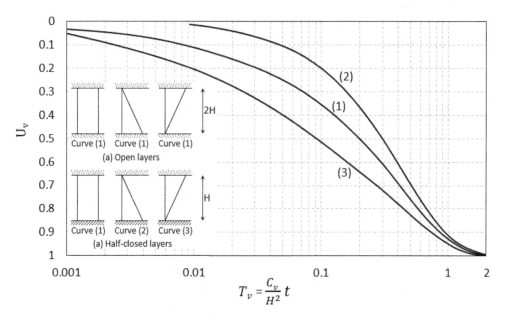

Figure 15.2 Relationship between the average degree of consolidation and time factor (modified after Knappett and Craig, 2012).

The equation for radial drainage only is:

$$\frac{\partial \bar{u}}{\partial t} = C_h \left[\frac{\partial^2 u}{\partial r^2} + \frac{1}{r} \frac{\partial u}{\partial r} \right] \tag{15.5}$$

The overall degree of consolidation U is given by:

$$(1-U) = (1-U_v)(1-U_r) \tag{15.6}$$

where U_v is the average degree of vertical consolidation, related to the time factor T_v, and U_r is average degree of radial consolidation, related to the time factor T_r.

The relationship between U_v and the time factor T_v can be represented graphically, as illustrated in Figure 15.2. The time factor T_v and T_r are given by:

$$T_v = \frac{C_v}{H^2} t \tag{15.7}$$

$$T_r = \frac{C_h}{r_e^2} t \tag{15.8}$$

with H = drainage path and r_e = radius of influence of the drain (smallest drainage path).

15.2.2 Installation pattern

Drains can be installed in square or triangular patterns, as illustrated in Figure 15.3.
If the drain radius is r_w, then

$$n = \frac{r_e}{r_w} \tag{15.9}$$

The spacing between two consecutive drains can be determined using the Barron curves (Fig.15.4).

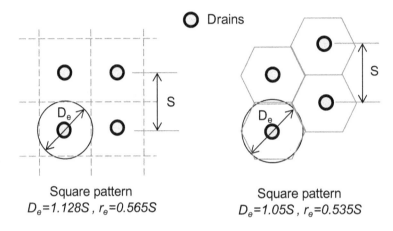

Figure 15.3 Typical drain installation patterns and equivalent diameters (modified after Indraratna et al., 2005); S = drain spacing.

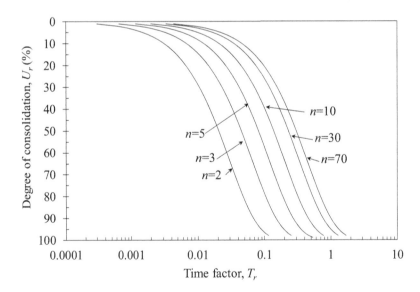

Figure 15.4 Relationship between U_r and T_r, e.g. Barron's chart (modified after Vinod et al., 2010).

The equivalent radius of a band drain is given by:

$$r_w = \frac{a+b}{4}$$

(15.10)

and a and b are the width and thickness of the drain, respectively.

15.2.3 Drain spacing (Rujikiatkamjorn and Indraratna, 2007, method)

The drain spacing is typically computed using dimensionless horizontal time factor-consolidation curves (T_r vs. U_r) charts (Fig. 15.4) as a function of n (i.e. Barron, 1948). Usually, the spacing is obtained via an iterative process to evaluate the required parameters such as n. As the manufacturer limits the availability of the size of PVDs, the appropriate design charts should be re-established using the equivalent drain diameter ($D_w = 2 r_w$) as a known variable in order to determine the drain spacing (S). This method can also capture the effect of the smear zone, and that is incorporated in the determination of parameter ξ as follows:

$$\xi = \left(\frac{k_h}{k_s}\right) \ln\left(\frac{D_s}{D_w}\right)$$

(15.11)

Where k_h is the coefficient of horizontal permeability, k_s is the coefficient of horizontal permeability in the smear zone, D_s is the diameter of the smear zone and D_w is the equivalent drain diameter.

In addition, Rujikiatkamjorn and Indraratna (2007) proposed empirical relationships for n based on the value of the parameter ξ as follows:

$$n = \exp\left(\alpha \ln \gamma + \beta\right)$$
$$\alpha = 0.3938 - 9.505 \times 10^{-4} \xi^{1.5} + 0.03714 \xi^{0.5}$$
$$\beta = 0.4203 + 1.456 \times 10^{-3} \xi^2 - 0.5233 \xi^{0.5}$$
$$\gamma = \frac{8 T'_h}{\ln \dfrac{(1-U)}{u^*}}$$

(15.12)

Where T'_h is the modified time factor to account for drain diameter ($= c_h\, t/D_w^2$). This is represented graphically in Figures 15.5 and 15.6.

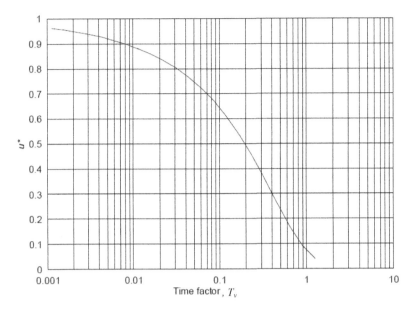

Figure 15.5 Average excess pore pressure due to vertical loading versus time factor T_v (Rujikiatkamjorn and Indraratna, 2007, reproduced with permission from *Canadian Geotechnical Journal*).

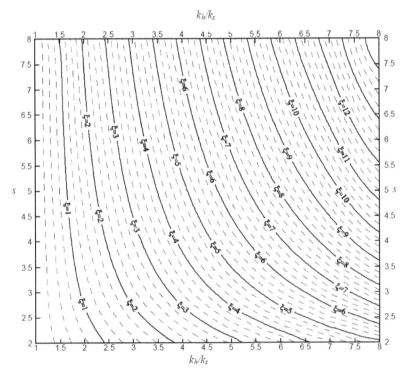

Figure 15.6 Design chart for the determination of the parameter ξ (after Rujikiatkamjorn and Indraratna, 2007, reproduced with permission from *Canadian Geotechnical Journal*).

Worked-out example 15.1: A basement of a building requires a 60×60 m excavation 2 m deep in a bed of sand 25 m thick. The minimum drawdown expected at the center of the excavation is 3 m, and permeability of sand is $k_{sand} = 4.8 \times 10^{-4}$ m/s. Prior to the dewatering system being installed, the groundwater level (GWL) was found to be at ground surface. The proposed dewatering system is to be set back from the excavation 5 m, with each well having a nominal diameter of 200 mm and pump operating capacity of 3000 L/min.

a. Determine the drawdown level expected at the wells.
b. Select the most appropriate well system.
c. Determine the number of wells and spacing (assume $C = 2,000$ (m/s)$^{-1/2}$).

Solution:
a. The profile of the ground is shown in Figure 15.7.

For an unconfined aquifer, we have:

$$q = \frac{k\pi \left(H_2^2 - H_1^2 \right)}{In\left(\dfrac{R_2}{R_1} \right)}$$

To calculate the expected drawdown at the wells, we can use the previous relationship, considering $R_1 = 0.1$ m (radius of the well) and $R_2 = 35$ m (center of the excavation) and $h_2 = (25-3) = 22$ m (center of excavation) and pump capacity of 3000 L/min or 0.05 m³/s.

$$0.05 = \frac{4.8 \times 10^{-4} \pi \left(22^2 - H_1^2 \right)}{In\left(\dfrac{35}{0.1} \right)}$$

$H_1 = 17.01$ m

Therefore, expected drawdown at the wells is $25-17.02 = 7.98$ m.

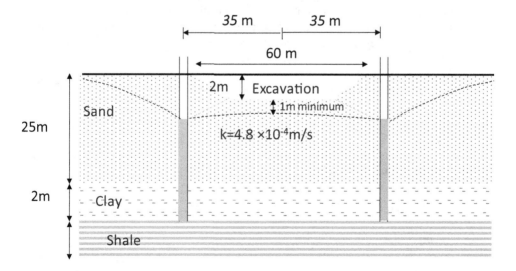

Figure 15.7 Ground profile for worked-out example 15.1.

Using Figure 15.1 (reproduced as Fig. 15.8), based on the expected drawdown and ground permeability, the most suitable dewatering system is the use of two stage well-points or deep wells.

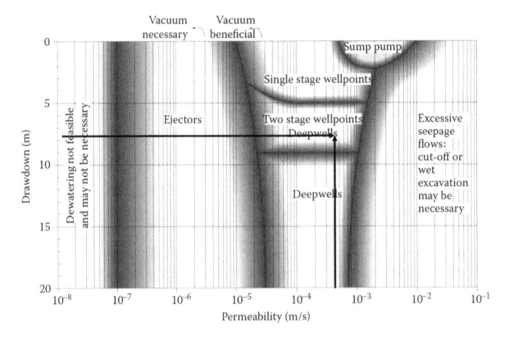

Figure 15.8 Range of application of dewatering techniques (after Powrie, 2014, reproduced with permission from CRC Press).

For the design of the dewatering system (no. of wells and spacing), the total amount of water flow that needs to be pumped out from the excavation has to be computed.

For this, we can consider the excavation as a single well with an equivalent radius:

$$B \times L = \pi R_w^2 \text{ or } R_w = \sqrt{\frac{B \times L}{\pi}}$$

For this excavation (60×60 m) with wells set back 5 m away, then:

$$R_w = \sqrt{\frac{70 \times 70}{\pi}} = 39.5 \text{ m}$$

For an unconfined aquifer we have:

$$q = \frac{k\pi \left(H_2^2 - H_1^2 \right)}{ln\left(\dfrac{R_2}{R_1} \right)}$$

In this problem, we can use the previous relationship, considering $R_w = R_1$ and radius of influence of the pumped zone = R_2 and $H_1 = (25-3) = 22$ m (center of excavation) and $H_2 = 25$ m (GWL at ground surface).

To compute the radius of influence of the pumped zone, Sichardt's formula can be used (minimum drawdown (H) at the center of the excavation = 3 m).

$$R_0 = C \times H \times \sqrt{k} = 2000 \times 3 \times \sqrt{4.8 \times 10^{-4}} = 131.5 \text{ m}$$

Then, substituting $R_0 = R_2$, the total amount of water flow can be computed as follows:

$$q = \frac{4.8 \times 10^{-4} \pi \left(25^2 - 22^2\right)}{ln\left(\dfrac{131.5}{39.5}\right)} = 0.177 \text{ m}^3/\text{s}$$

The principle of superposition can be used to calculate the combined drawdown of multiple wells.

If each well pump has an operating capacity of 3000 L/min or 0.05 m^3/s, then the required number of wells can be calculated as follows:

No. of wells = 0.177 / 0.05 = 3.54 wells
The number of wells required is 4.

Well location on the excavation perimeter is 70 × 4 = 280 m, then well spacing is approximately 280/4 = 70 m.

Worked-out example 15.2: A new extension for a port infrastructure is proposed. To reclaim the land for the new extension, a 2.5 m thick layer of hydraulic fill ($\gamma_{sat} = 15$ kN/m^3) is placed over a 6 m thick deposit of normally consolidated clay, resting on impermeable shale (Fig. 15.9). To accelerate drainage and consolidation of the clay layer, it is proposed to install a series of radial sand drains in sufficient number so that the time needed for 90% overall consolidation to occur is reduced to just 2 years. The sand drains to be installed are 200 mm in diameter and will be inserted throughout the clay layer using the triangular pattern. Laboratory tests revealed that the clay has a coefficient for vertical and horizontal consolidation of $c_v = 2.2$ m^2/year and $c_h = 3$ m^2/year, respectively.

Calculate the spacing between two consecutive drains needed to achieve the stated objective.

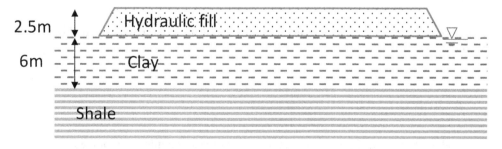

Figure 15.9 Profile for worked-out example 15.2.

Solution: Calculate the time factor for vertical drainage only.

$$T_v = \frac{C_v}{H^2}t = \frac{2.2 \times 2}{6^2} = 0.122$$

Then calculate the degree of consolidation due to vertical drainage only for the 2-year period using the Terzaghi chart (Fig. 15.3) reproduced from U_v versus T_v (Fig. 15.10).

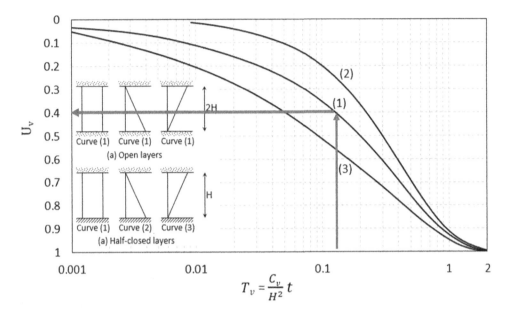

Figure 15.10 Relationship between the average degree of consolidation and time factor (modified after Knappett and Craig, 2012).

From the graph:

$$U_v \approx 0.40$$

As the required overall degree of consolidation for 2 years is $U = 90\%$:

$$(1-U) = (1-U_v)(1-U_r)$$

Therefore, in order to have an overall consolidation of 90%, the required radial degree of consolidation U_r for 2 years is:

$$(1-0.9) = (1-0.4)(1-U_r)$$

$$U_r = 0.83$$

The sand drain must be designed so that the system reaches a minimum of 0.83 degrees of consolidation due to radial drainage.

To determine the drain spacing required for reaching the degree of consolidation of 0.83, an iterative process between an initial assumed n compared to an obtained n must be used.

Use Barron charts (Fig. 15.4, reproduced in Fig. 15.11) for radial consolidation.

Iteration 1: For $U_r = 0.83$ and $n_1 = 10$, from the graph (Fig. 15.4, reproduced in Fig. 15.11, noting the relevant U_r), $T_r \approx 0.34$ (as drain diameter is 200 mm, $r_w = 0.1$ m).

$$T_r = \frac{C_h}{4 r_e^2} t \text{ solving for } r_e \text{ we have } r_e = \left[\frac{C_h}{4 T_r} t \right]^{1/2} = \left[\frac{3}{4 \times 0.34} \times 2 \right]^{1/2} = 2.1$$

For this calculated r_e, a new value of $n = n_2$ can be found:

$$n_2 = \frac{r_e}{r_w} = \frac{1}{r_w} \left[\frac{C_h}{4 T_r} t \right]^{1/2} = \frac{2.1}{0.1} = 21$$

$$n_1 = 10 \neq n_2 = 21$$

Iteration 2: A new n is assumed, $n_3 = 15$ can be considered.

Using the Barron chart for $U_r = 0.83$ and $n_3 = 15$, then $T_r \approx 0.43$.

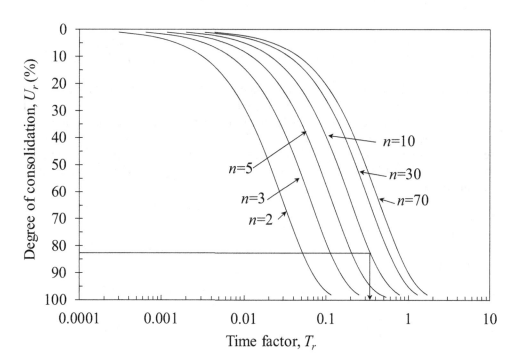

Figure 15.11 Relationship between U_r and T_r (modified after Vinod et al., 2010).

A n value of n_4 is:

$$n_4 = \frac{r_e}{r_w} = \frac{1}{r_w}\left[\frac{C_h}{4T_r}t\right]^{1/2} = \frac{1}{0.1}\left[\frac{3}{4\times0.43}\times2\right]^{\frac{1}{2}} = 18.7$$

$n_3 \neq n_4$ (i.e. different is too high)

Iteration 3: A new n is assumed, $n_5 = 17$ can be considered.
    Using the Barron chart for $U_r = 0.83$ and $n_5 = 17$, then $T_r \approx 0.46$.
    A new n value of n_6 is:

$$n_6 = \frac{r_e}{r_w} = \frac{1}{r_w}\left[\frac{C_h}{4T_r}t\right]^{1/2} = \frac{1}{0.1}\left[\frac{3}{4\times0.46}\times2\right]^{\frac{1}{2}} = 18$$

$n_6 \approx n_5$

Since the drains are installed in a triangular pattern, then:

$$n_6 = \frac{r_e}{r_w} = 18$$

$$r_e = 18\times0.1 = 1.8\,\text{m}$$

$$R = r_e = 1.8 = 0.525\times S$$

$$S = 3.4\,\text{m}$$

For improved efficiency and added reliability, keep the spacing of sand drains at 3 m.

Worked-out example 15.3: A road embankment is planned to be constructed over a layer of saturated clay of 8 m thick. To increase the rate of consolidation, it is advised to install 350 mm diameter sand drains at 3 m spacing in a square pattern. The degree of consolidation, within the time due for operation, was estimated to be 30% without the drains. Calculate the degree of consolidation of the clay for the same time after the installation of drains. Consider an impermeable layer below the clay and $c_v = 4.0$ m²/yr and $c_h = 6.9$ m²/yr. Ignore the stiffness of the sand drains.

Solution:
    Without any sand drains:

$$U_V = 0.3$$

From Fig. 15.2: $T_v = 0.071$

$$t = \frac{T_v d^2}{c_v} = \frac{0.071\times8^2}{4} \text{ or } t = 1.14 \text{ years}$$

After the installation of the drains (square pattern):

$$r_e = 0.564 \times s = 0.564 \times 3 = 1.69\,\text{m}$$

$$n = \frac{r_e}{r_d} = \frac{1.69}{0.175} = 10$$

$$T_r = \frac{c_h}{4r_e^2} \times t = \frac{6.9 \times 1.14}{4 \times 1.69^2} = 0.7$$

From Fig. 15.4: $U_r = 0.97$

$$(1-U) = (1-U_v)(1-U_v)$$
$$U = 1-(1-0.3)(1-0.97)$$
$$U = 98\%$$

Worked-out example 15.4: An embankment is to be constructed as a foundation of a road-way, on top of the 8.0 m thick layer of firm clay, sandwiched between silty sand at the top and dense sand at the bottom.

To accelerate drainage, it is proposed to install a series of prefabricated vertical drains (PVDs) in sufficient number so that the time needed for 95% overall consolidation to occur is reduced to just 1 year. The PVDs to be installed have an equivalent diameter of 50 mm (based on Eq. 15.10) and will be inserted throughout the clay layer using the triangular pattern. The coefficients of horizontal and vertical consolidation are $c_h = 0.288$ m²/month, $c_v = 0.187$ m²/month.

A preliminary drain installation trial to evaluate the effect of smear caused by the mandrel during the installation of the PVDs revealed that the ratio of horizontal permeability in the smear zone (k_h/k_s) is 5, whereas the diameter of the smear zone smear zone (D_s) is found to be 0.15 m.

Calculate the spacing between two consecutive drains needed to achieve the stated objective.

Solution: As the Rujikiatkamjorn and Indraratna (2007) method (Fig. 15.12) has a provision for considering the effect of smear, this method is preferred.

Considering the permeability ratio and the diameter of the smear zone given, we have:

$$\xi = \left(\frac{k_h}{k_s} - 1\right)\ln\left(\frac{D_s}{D_w}\right) = (5-1)\ln\left(\frac{0.15}{0.05}\right) = 4.4$$

For $\xi = 4.4$:
then:

$$\alpha = 0.3938 - 9.505 \times 10^{-4}\xi^{1.5} + 0.03714\xi^{0.5} = 0.463$$
$$\beta = 0.4203 + 1.456 \times 10^{-3}\xi^{2} - 0.5233\xi^{0.5} = -0.649$$

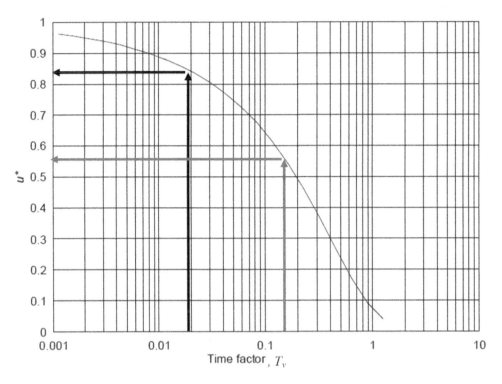

Figure 15.12 Average excess pore pressure due to vertical loading versus time factor T_v (Rujikiatkamjorn and Indraratna, 2007, reproduced with permission from *Canadian Geotechnical Journal*).

The average excess pore pressure due to vertical loading (u^*) can be computed using Figure 15.12 and considering $T_v = 0.14$ (from the equation that follows).

$$T_v = \frac{C_v}{H^2} t = \frac{0.187 \times 12}{4^2} = 0.14$$

Then, $u^* = 0.56$ (Fig. 15.12, grey arrows).
    Using the data given,

$$T'_h = \frac{c_h t}{D_w^2} = \frac{0.288 \times 12}{(0.05)^2} = 1382.4$$

$$\gamma = -\frac{8 T'_h}{\ln\left(\frac{1-U}{u^*}\right)} = -\frac{8 \times 1382.4}{\ln\left(\frac{1-0.95}{0.56}\right)} = 4577.65$$

And

$$n = \exp(\alpha \ln \gamma + \beta) = \exp(0.463 \ln(4577.65) - 0.649) = 25.8$$

Then, $n = \dfrac{D_e}{D_w}$, or $D_e = 25.8 \times 0.05 = 1.29$ m.

For a triangular pattern, $D_e = 1.05\, S$, thus drain spacing (S) is:

$S = 1.29/1.05 = \; = 1.22 \approx 1.2$ m.

Worked-out example 15.5: A 3 m embankment ($\gamma = 20$ kN/m^3) is to be constructed as a foundation of a roadway, on top of the 8.0 m thick layer of firm clay, sandwiched between a silty sand at the top and an impermeable rock layer at the bottom.

To accelerate drainage, it is proposed to install a series of prefabricated vertical drains (PVDs) in sufficient number with a vacuum preloading of 60 kPa so that the time needed for 95% overall consolidation to occur is reduced to just 6 months. The PVDs (100 mm wide band drains) to be installed have an equivalent diameter of 50 mm (see Eq. 15.10) and will be inserted throughout the clay layer using the triangular pattern. The coefficients of horizontal and vertical consolidation are $c_h = 0.288$ m^2/month, $c_v = 0.187$ m^2/month

A preliminary drain installation trial to evaluate the effect of smear caused by the mandrel during the installation of the PVDs revealed that the ratio of horizontal permeability in the smear zone (k_h/k_s) is 5, whereas the diameter of the smear zone smear zone (D_s) is found to be 0.15 m.

Calculate the spacing between two consecutive drains needed to achieve the stated objective.

Solution: Based on Example 15.4, the following parameters can be obtained:

$\xi = 44$
$\alpha = 0.463$
$\beta = -0.649$

The average excess pore pressure due to vertical loading (u^*) can be computed using Fig.15.12 and considering $T_v = 0.02$ (from the calculation below).

$$T_v = \frac{C_v t}{H^2} = \frac{0.187 \times 6}{8^2} = 0.02$$

Then, $u^* = 0.83$ (Fig. 15.12).

For vacuum pressure and surcharge preloading, the general equation for γ is modified as follows:

$$\gamma = -\frac{8T'_h}{\ln\left(\dfrac{1-U_{t,vac}}{u^*}\right)}$$

where $U_{t,vac}$ is the equivalent degree of consolidation incorporating the combined effect of vacuum pressure (u_{vac}) and the vertical surcharge load (u_0).

The vertical surcharge load can be computed considering the unit weight of the embankment fill as follows:

$$u_0 = 20 \times 3 = 60\,\text{kPa}$$

Then,

$$U_{t,vac} = \frac{U_t \times u_0}{u_0 + |u_{vac}|} = \frac{0.95 \times 60}{60 + 60} = 0.48$$

The numerator of the previous equation assumes that the maximum pore pressure for this case will be obtained by multiplying the expected degree of consolidation and maximum surcharge load. Furthermore, in this approach, it is assumed the surcharge pressure is applied all at once, which does not reflect construction practice where surcharge load is increased incrementally.

Then,

$$T_h' = \frac{c_h t}{D_w^2} = \frac{0.288 \times 6}{(0.05)^2} = 691.2$$

$$\gamma = -\frac{8T'_h}{\ln\left(\frac{1 - U_{t,vac}}{u^*}\right)} = -\frac{8 \times 691.2}{\ln\left(\frac{1 - 0.48}{0.83}\right)} = 11825.57$$

and

$$n = \exp(\alpha \ln \gamma + \beta) = \exp(0.463 \ln(11825.57) - 0.649) = 40$$

Then, $n = \dfrac{D_e}{D_w}$, or $D_e = 40 \times 0.05 = 2$ m.

For a triangular pattern, $D_e = 1.05\,S$, thus drain spacing (S) is:

$$S = 2/1.05 = 1.92 \approx 1.9 \text{ m}$$

The value for drain spacing obtained considering vacuum pressure is larger compared with the case considering surcharge only, which clarifies the role of suction effecting consolidation.

In practical situations, considering possible vacuum loss in the ground, the drain spacing can then be adjusted to 1.5 m.

References

Barron, R.A. (1948) Consolidation of fine grained soils by drain wells. *Transactions of the American Society of Civil Engineers*, 113, 718–724.

Indraratna, B., Chu, J. and Rujikiatkamjorn, C. (2005) *Ground Improvement: Case Histories*. Elsevier Science, 608p. ISBN: 9780080446332.

Knappett, J. & Craig, R.F. (2012) *Craig's Soil Mechanics*, 8th edition. CRC Press, ISBN:9780415561266, p. 584.

Powrie, W. (2014) *Soil Mechanics – Concepts and Applications*, 3rd edition. CRC Press, Taylor & Francis, United Kingdom, ISBN-10: 041531156X.

Rujikiatkamjorn, C. & Indraratna, B. (2007) Analytical solutions and design curves for vacuum-assisted consolidation with both vertical and horizontal drainage. *Canadian Geotechnical Journal*, 44(2), 188–200. https://doi.org/10.1139/t06-111.

Sichardt, W. (1928) *Das Fassungsvermogen von Rohrbrunnen und seine Bedeutung fur die Grundwasserabsenkung, insbesondere für größere Absenkungstiefen.* Springer, Berlin.

Vinod, J.S., Sridharan, A. & Indraratna, B. (2010) Determination of coefficient of radial consolidation using steepest tangent fitting method. *Geotechnical and Geological Engineering Journal,* 28(4), 533–534.

Index

Note: Page numbers in **bold** indicate a table and page numbers in *italics* indicate a figure on the corresponding page.